Primate Phylogeny

Edited by
F. E. GRINE, J. G. FLEAGLE and L. B. MARTIN
Department of Anthropology,
State University of New York,
Stony Brook, NY 11794, U.S.A.

ACADEMIC PRESS
Harcourt Brace Jovanovich, Publishers
London San Diego New York Boston
Sydney Tokyo Toronto

ACADEMIC PRESS LIMITED
24–28 Oval Road
London NW1 7DX
(Registered Office)

United States edition published by
ACADEMIC PRESS INC.
San Diego, CA 92101

(Reprinted from *Journal of Human Evolution*, Vol. 16, No. 1, 1987)

ISBN 0-12-303960-6

Printed in Great Britain by Black Bear Press Limited, Cambridge

List of Contributors

Dr P. Andrews, *Department of Palaeontology, British Museum (Natural History), Cromwell Road, London SW7 5BD, England*

Dr A. T. Chamberlain, *Hominid Palaeontology Research Group, Department of Human Anatomy and Cell Biology, The University of Liverpool, PO Box 147, Liverpool L69 3BX, England*

Dr H. H. Covert, *Department of Anthropology, The University of Colorado, Boulder, Colorado 80309-0233, U.S.A.*

Dr E. Delson, *Department of Anthropology, Lehman College and the Graduate School, CUNY; and the Department of Vertebrate Paleontology, American Museum of Natural History, New York, NY 10024, U.S.A.*

Dr T. Harrison, *Department of Anthropology, New York University, 25 Waverly Place, New York, NY 10003, U.S.A.*

Dr L. B. Martin, *Department of Anthropology, State University of New York, Stony Brook, NY 11794, U.S.A.*

Dr J. H. Schwarz, *Department of Anthropology, University of Pittsburgh, Pittsburgh, PA 15260, U.S.A.*

Dr E. Strasser, *Department of Anthropology, Graduate School, City University of New York, NY 10036, U.S.A.*

Dr C. B. Stringer, *Department of Palaeontology, British Museum (Natural History), Cromwell Road, London SW7 5BD, England*

Dr I. Tattersall, *Department of Anthropology, American Museum of Natural History, New York, NY 10024, U.S.A.*

Dr J. R. Wible, *Department of Anatomy, The University of Chicago, Chicago, Illinois 60637, U.S.A.*

Dr B. A. Wood, *Hominid Palaeontology Research Group, Department of Human Anatomy and Cell Biology, The University of Liverpool, PO Box 147, Liverpool L69 3BX, England*

Preface

The papers that compose this volume derive from a symposium on Primate Phylogeny that was held at the American Museum of Natural History, New York, 9 November, 1986 during the course of the Sixth Annual Meeting of the Willi Hennig Society. The symposium was organized at the request of Steve Farris for the Council of the Willi Hennig Society, as the first of an anticipated series through which the application of cladistic techniques to various groups of organisms might be encouraged. The session was well attended and the papers presented during the course of the afternoon generated considerable discussion that was curtailed only by the need to close the auditorium for the evening. Comments on and impressions of the symposium have been recorded by both Bernor (1986) and Coddington (1987).

In organizing the symposium and this volume, we attempted to bring together scientists who are actively engaged in studies of a wide range of living and fossil primates. Each of the contributors was requested to address the major issues concerning the phylogeny of a subset of the Order Primates. The contributors were also encouraged to tabulate the distributions of the morphological features upon which their analyses and conclusions were based, as we had hoped originally to combine all of the data from the individual papers into a single quantitative analysis of primate phylogeny for inclusion in this volume. This, unfortunately, proved an impossible undertaking for the present.

Perhaps the most surprising outcome of the symposium was the extraordinary heterogeneity of approaches that are currently being pursued under the general umbrella of phylogenetic analysis. Students of primate phylogeny, like those investigating the evolutionary relationships of other groups of organisms, are in considerable disagreement regarding such basic issues as what types of data are most appropriate for understanding phylogeny, whether characters should be weighted, how character polarity should best be determined, and even whether parsimony is a suitable criterion for evaluating the relative merits of competing phylogenies.

In general, however, the papers in this volume are concerned less with specific issues of methodology than with the more general issue of constructing phylogenies on the basis of shared-derived characters rather than on general phenetic similarities, geographical proximity, or less explicit criteria such as tradition. This approach is by no means universally acknowledged or followed by students of primate evolution, and the papers in this issue will hopefully serve to focus efforts at phylogeny reconstruction along cladistic lines among primatologists, and to form the basis for more detailed, quantitative studies in the future.

Because of time and page constraints, many of the contributors have limited themselves to a presentation of what they consider to be the most likely phylogeny for the group under consideration, and to a summary of the derived features that characterize each node.

Primates have been studied quite intensively by comparison with most other vertebrate orders. While there is relatively little disagreement regarding the phylogeny of some groups (e.g., Old World monkeys), for other groups (e.g., "prosimians" and fossil hominids) the phylogenetic relationships are considerably more ambiguous, and in these instances the contributors compare data that have been advanced in support of alternate and often radically different phylogenies. Most of the primate groups compose clades that appear to be unambiguously supported by large numbers of derived features, as well as taxa whose relationships are subject to conflicting arrangements. Thus, most of the contributions to this volume present a combination of more secure and less secure hypotheses of relationship that deserve critical evaluation and testing.

F. E. Grine
J. G. Fleagle
L. B. Martin
Stony Brook,
September 1987

References

Bernor, R. L. (1986). Symposium on primate phylogeny: Willi Hennig Society, 9 November 1986. *J. hum. Evol.* **15**, 409–412.

Coddington, J. A. (1987). The sixth annual meeting of the Willi Hennig Society. *Cladistics* **3**, 178–185.

CONTENTS

John R. Wible

Department of Anatomy, The University of Chicago, Chicago, Illinois 60637, U.S.A.

Herbert H. Covert

Department of Anthropology, The University of Colorado, Boulder, Colorado 80309-0233, U.S.A.

Received 13 February 1987
Revision received 27 May 1987
and accepted 28 May 1987

Publication date November 1987

Keywords: Primates, Euprimates, archaic primates, Archonta, tree shrews, cladistic analysis, diagnosis.

Primates: cladistic diagnosis and relationships

The morphological evidence for the phylogenetic relationships of euprimates, archaic primates, and related eutherian orders is reviewed following the methods of Hennigian phylogenetic systematics. Euprimates, the group including living primates and their closest common ancestor, is diagnosed by a suite of shared derived characters of the cranium and postcranium exhibiting relatively unique distributions among Eutheria. Plesiadapiformes, the group of archaic primates generally held to be the sister group to Euprimates, is not demonstrably monophyletic (with or without Microsyopidae). The Superorder Archonta (primates, tree shrews, bats, and colugos) is the only higher-level grouping including Euprimates that is based on uniquely derived morphological characters. Hypotheses of relationships within Archonta ally Euprimates with either tree shrews or some plesiadapiforms (paromomyids and plesiadapids), but the euprimate-tree shrew clade receives more support from the distribution of derived characters among the taxa studied. Because the higher-level affinities of Euprimates are not well resolved, we advocate equating the Order Primates with the taxon Euprimates.

Journal of Human Evolution (1987) **16,** 1–22

Introduction

The past twenty years have witnessed a remarkable rebirth of interest in the superordinal relationships within Mammalia (Van Valen, 1971; McKenna, 1975; Szalay, 1977; Novacek, 1982, 1986). The reasons for this rebirth are many and varied. Included are the general increase of interest in phylogenetic systematics following the work of Hennig (1966), the marked increase in our knowledge of the Mesozoic and Cenozoic fossil records (Lillegraven *et al.*, 1979; Savage & Russell, 1983), and the appearance of numerous molecular techniques for assessing phylogeny (Goodman, 1982). Mammalian systematists are, as yet, far from agreement on most aspects of the branching pattern of the mammalian ordinal tree, but some points of congruence have been reached. For one, there is widespread agreement that lagomorphs and rodents should be united in the Superorder Glires (Luckett & Hartenberger, 1985).

Throughout this rebirth, the Order Primates has remained a probem for morphologists interested in mammalian superordinal relationships. There is no consensus about what the order comprises, and this presents problems for analyses of higher-level relationships. The current state of affairs is, in large measure, a historical holdover. In a series of very influential publications culminating in the 1959 book *The Antecedents of Man*, Le Gros Clark promoted the view that the Order Primates is best defined in terms of a suite of evolutionary trends, rather than clear-cut, identifiable shared unique specializations. These trends, as seen from modern tree shrews to man, include, for instance, the progressive abbreviation of the snout with reduction of the apparatus of smell, and the progressive expansion and elaboration of the brain. A definition of Primates based on progressive changes in features lent support to the step-wise gradistic view of primate evolution also promoted by Le Gros Clark: that *en route* to man, primates went through a tree shrew-like stage, a lemur-like stage, a monkey-like stage, etc. The trends and grades of

0047–2484/87/010001 + 22 $03.00/0

Le Gros Clark's writings have greatly influenced the current generation of primatologists, many of whom continue to define the order in terms of adaptive grades rather than evolutionary clades (MacPhee *et al.*, 1983; Gingerich, 1986; see below).

Attempts have been made to delineate Primates using the methodology of Hennigian phylogenetic systematics. The first was by Martin (1968). He concluded, as Van Valen (1965) had several years earlier, that the features shared by tupaiids (tree shrews) and primates are either primitive retentions (plesiomorphies) or independent acquisitions and, as such, cannot be used to unite these groups in the same order. With the removal of tree shrews from the order, Martin went on to identify five shared derived characters (synapomorphies) for Primates. Only two of these characters involve hard tissues identifiable in fossil forms: the inclusion of a plate of the ethmoid in the wall of the orbit and an auditory bulla formed from the petrosal bone. Martin offered this list of synapomorphies more as a program for future research than as a definition of the order. Nonetheless, this "definition" was not very satisfactory for the paleontologist attempting to delineate the boundaries of the primate order. Cranial material was required for analysis, while the bulk of the Paleogene fossil record was teeth.

Recently, Martin (1986) has updated his 1968 article to provide a much more detailed definition of the Order Primates. He specifically tackles the problem of delineating the primate boundary in the fossil record. His definition of modern primates includes six hard-tissue features that are applicable to analyses of fossil remains. These are (p. 20):

(i) Well-developed, divergent hallux with a flat terminal phalanx in the foot.
(ii) Elongated distal segment of the calcaneus (viz. low calcaneal index value).
(iii) Relatively large, convergent orbits with restricted interorbital distance. Postorbital bar present; ethmoid exposure in the orbit possible (depending on interorbital distance relative to skull size).
(iv) Petrosal bulla.
(v) Relatively large braincase. Sylvian sulcus on endocast.
(vi) Dental formula maximally

$$\frac{2\,.\,1\,.\,3\,.\,3\,.}{2\,.\,1\,.\,3\,.\,3\,.}.$$

Premaxilla short; upper incisors arranged more transversely than longitudinally. Molars with low, rounded cusps. Lower molars with raised, enlarged talonids.

Martin noted that this list is not without problems when applied to the primate fossil record. For example, a number of Eocene mammals that are clearly primates on other grounds retain four premolars above and below. We add that the molar features he identifies as synapomorphies of modern primates are so vague and characteristic of so many mammals that they are useless for delineating primates from non-primates in Paleogene collections. Likewise, a feature such as a relatively large braincase, while most certainly true, is problematic for the paleontologist, because so few relatively complete Paleogene mammalian crania are known. These problems notwithstanding, Martin concluded that the Late Cretaceous-Paleogene plesiadapiforms, the so-called archaic primates, should be excluded from the Order Primates, because there is no evidence that these forms had any of the above listed six hard-tissue primate features.

The years between the publication of Martin's two contributions have witnessed a very lively debate about the delineation of the Order Primates. Controversy has concerned not only discussions of specific taxa and features, but also discussions of the criteria for defining orders. Cartmill (1972, 1974) argued that ordinal boundaries should reflect major adaptive shifts, that without this criterion the "placement of any taxonomic boundary between ancestral and descendant groups becomes largely a matter of caprice, subject only to the criterion of monophyly" (1972, p. 98). Szalay and co-workers (Szalay & Decker, 1974; Szalay *et al.*, 1975) followed a similar approach, identifying as primate synapomorphies not only specific postcranial specializations, but also "the basic adaptations to an arboreal habitat from an ancestry that was probably habitually terrestrial" (Szalay *et al.*, 1975, p. 165). Yet, the primate boundary and adaptive shift proposed by these authors differed from that postulated by Cartmill (1972, 1974). Szalay and co-workers included the Plesiadapiformes (*sensu* Simons & Tattersall in Simons, 1972) within the order because they shared postcranial specializations indicative of an arboreal adaptation with other primates; Cartmill excluded that group and proposed that "the ancestral primate adaptation involved nocturnal, visually-directed predation on insects in the terminal branches of the lower strata of tropical forests" (1972, p. 121).

Some primatologists have returned to a gradistic approach to classification like that promoted for primates by Le Gros Clark. MacPhee *et al.* (1983) identified three primate grades linked by unresolved phylogenetic relationships: a plesiadapiform grade (including Plesiadapidae, Carpolestidae, Paromomyidae, Picrodontidae, Microsyopidae, Apatemyidae, Mixodectidae, Tupaiidae, and Dermoptera), a prosimian grade (including Tarsiidae, Lemuriformes, Lorisiformes, Omomyidae, and Adapidae), and an anthropoid grade (including Anthropoidea). They concluded that:

> The search for exclusive features diagnostic for primates might be better replaced by attempts to formulate and account for the significant grade boundaries found within the order. Only two major grade transitions, delimiting the upper and lower boundaries of Prosimii, can be defined at present. The antecedent plesiadapiform grade cannot be precisely defined, and it may even be polyphyletic. Polyphyletic origin of a grade damages its taxonomic status but enhances its evolutionary significance, because a grade boundary crossed by several parallel lineages is correspondingly more likely to reflect adaptation under pervasive selection pressures rather than historical accident (p. 511).

More recently, Gingerich (1986) argued that a gradistic approach to delineating primates is more informative than one based on clades. "A grade classification appears to express what we know (and do not know) about the phylogeny of early primates more accurately than a cladistic classification" (1986, p. 40).

As this survey of the literature reveals, vastly different opinions have been expressed recently regarding the delineation of the Order Primates. At the extremes are the gradistic approach advocated by MacPhee *et al.* (1983) and Gingerich (1986) and the Hennigian approach of Martin (1986). The former essentially argues that we are unable to delineate the primate order on the basis of a suite of shared derived characters and that it is unlikely that we will be able to do so in the future. These authors figure a history of Primates that includes not only Plesiadapiformes (*sensu* Simons & Tattersall in Simons, 1972), but also Microsyopidae, Apatemyidae, Mixodectidae, Tupaiidae, and Dermoptera. In contrast, Martin argues that we can and should offer a concise definition for modern primates and that fossil forms which fit (or nearly fit) this definition be considered primates. Martin explicitly excludes the Plesiadapiformes and Tupaiidae from the Order Primates. In the following, we review the evidence upon which these hypotheses are based, that is, the evidence behind the various hypotheses of phylogenetic affinities between modern and archaic primates and related eutherian groups. In this, we choose to follow the Hennigian (cladistic) approach for reasons enumerated below.

Methods and materials

According to Hennig (1966), the basis for the classification of organisms is genealogical relationship or phylogeny. Such relationships are recognized through the sharing of evolutionary novelties (derived or apomorphous characters) inherited through a common ancestor and are usually depicted in the form of branching diagrams or cladograms. Natural groups of organisms (monophyletic groups or clades) are recognized and arranged in hierarchical relationships to one another by the hierarchical distribution of shared derived characters (synapomorphies). Primitive (plesiomorphous) states of characters cannot be used to recognize natural groups because they originated as evolutionary novelties in an earlier ancestor and provide no information for subsequent branching points in the phylogenetic tree. Unnatural groups of organisms, those not recognized by synapomorphies alone, are termed paraphyletic or polyphyletic groups (for further discussion see Wiley, 1981).

The entire organism, including its development and mode of life, is the pool from which characters should be sampled. Yet, unavoidably in a review such as this, it is only those portions of the organism already somewhat well known that can be sampled. With regard to extinct and extant primates and related eutherian orders, the basicranium and pes have been the most intensively studied portions of the anatomy over the last years, and it is here that we concentrate. We rely heavily on published descriptions of these anatomical regions, in particular, for Paleogene primates, because we have not studied many of the relevant fossils. Polarity of characters, their primitive (plesiomorphous) and derived (apomorphous) states, is determined using the method of outgroup analysis (Maddison *et al.*, 1984), which assumes that the primitive state is that found in the closest relatives of the group under study. As a drawback, however, this method assumes that outgroup relationships are well resolved, and unfortunately the branching pattern of the eutherian ordinal tree is still somewhat obscure. We therefore feel compelled in allying groups to rely on those derived characters which have been demonstrated to exhibit little homoplasy among other eutherians (see Szalay & Drawhorn, 1980; Novacek, 1986; MacPhee & Cartmill, 1986), that is, derived characters which have a unique or relatively unique distribution among the taxa studied.

Cartmill (1974, p. 442) argues that "taxonomic boundaries must reflect more than merely phylogenetic affinity; they must also mark important adaptive shifts that underlie the evolutionary trends characteristic of a radiating higher taxon." Above, we have presented the example of the differing adaptive shifts postulated for the primates by Cartmill (1972, 1974) on the one hand and Szalay *et al.* (1975) on the other. This example indicates the subjectivity involved in choosing which features to analyze for the reconstruction of "adaptive shifts". We consider the delineation of taxonomic units and the adaptations of those units as separate questions (Lauder, 1981). One cannot ask questions about the initial adaptations of a group without knowing which organisms are included and which are excluded. The delineation of higher taxonomic groups that include fossil forms has long been problematic for systematists, in part, because there is disparity between our knowledge of modern and fossil organisms. Within Hennigian classification, two major approaches for naming higher taxa have surfaced. The discussion of these is facilitated by introducing the terms "crown group", "stem lineage", and "closed descent community" (Hennig, 1981; Ax, 1985) (Figure 1). A crown group constitutes the living members of a group and all fossil taxa thought to be descended from the closest common ancestor of the living members of the group. A stem lineage includes the fossil taxa more closely related to a particular crown group than to its living sister group. Finally, a closed descent community includes a crown group and its stem lineage. Some authors (Patterson & Rosen, 1977) equate the name of a living group with the closed descent community; others (Gauthier, 1986; Gauthier *et al.*, in press) restrict the name to the crown group. We will evaluate the results of our review in light of both approaches.

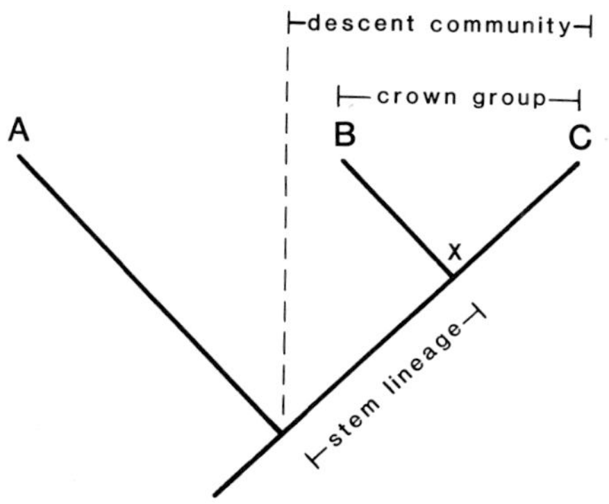

Figure 1. Cladogram illustrating the terms "crown group", "stem lineage", and "closed descent community" (Hennig, 1981; Ax, 1985). The crown group B–C includes the modern taxa B and C and their closest common ancestor, X. The stem lineage of the crown group B–C includes the outlying fossil taxa more closely related to B–C than to its modern sister taxon A. The closed descent community of B–C includes the crown group and stem lineage.

MacPhee *et al.* (1983) and Gingerich (1986) make various claims about the desirability of grades over clades in delimiting the primate boundary. Nowhere do these authors provide a rationale for delimiting grades within primates; at most, they admit that the plesiadapiform (praesimian of Gingerich) grade cannot be defined. What is the purpose of postulating a grade without any rationale for its limits? Moreover, we do not understand how a group of organisms for which no delimiting features are known can somehow reflect "pervasive selection pressures".

Although the term "definition" was employed throughout the introduction (for example, the definition of the Order Primates), our usage followed that of the authors discussed. "Definition" is a highly equivocal term and should be replaced with the term "diagnosis" in systematics (Ghiselin, 1984). "In conventional logic, a definition states the attributes necessary and sufficient for a name to apply" (*ibid.*, p. 105). We do not proceed by assigning necessary and sufficient attributes to define taxa in systematics, that is, we do not set the boundaries to define groups. We recognize natural groups through the methods discussed above, that is, by observing that they share derived characters. The term "diagnosis" applies to the "brief listing of those characters which differentiate a taxon from related and/or similar taxa" (Wiley, 1981, p. 377). Diagnostic properties do not define taxa in the conventional logic sense, but are useful for identification of taxa.

Results

Euprimates monophyly

The Euprimates of Hoffstetter (1977), the "primates of modern aspect" of Simons (1972), includes the modern lemurs, lorises, galagos, tarsiers, monkeys, apes, and man and a range of fossil forms, including the Eocene adapids and omomyids. Although contrary views have been proposed in the past (Gingerich, 1976; Schwartz *et al.*, 1978), current consensus (Martin, 1986; Gingerich, 1986) holds that Euprimates is a monophyletic group (Figure 2). Modern euprimates are distinguished from other mammals by a number of hard- and soft-tissue features (Martin, 1986), but most of the soft-tissue features cannot be identified

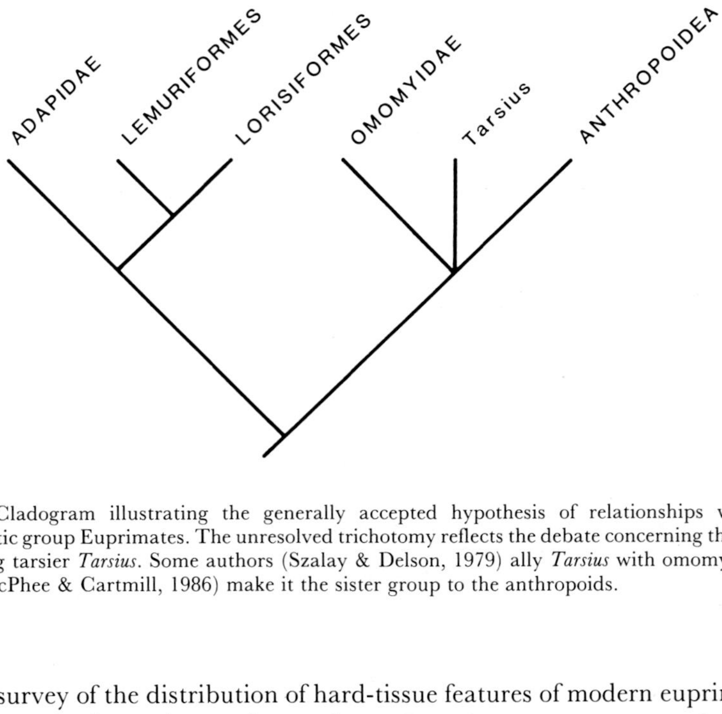

Figure 2. Cladogram illustrating the generally accepted hypothesis of relationships within the monophyletic group Euprimates. The unresolved trichotomy reflects the debate concerning the affinities of the living tarsier *Tarsius*. Some authors (Szalay & Delson, 1979) ally *Tarsius* with omomyids; while others (MacPhee & Cartmill, 1986) make it the sister group to the anthropoids.

in fossils. A survey of the distribution of hard-tissue features of modern euprimates across Mammalia reveals that only one feature is totally unique among mammals, while eight others are relatively unique (Table 1). Of these nine derived features of modern euprimates, eight are identifiable in representatives of the two Eocene euprimate groups, Adapidae and Omomyidae (see Szalay & Delson, 1979; MacPhee & Cartmill, 1986). The exception is the orbital exposure of the ethmoid bone, which has been found thus far in only one adapid, the Cambridge skull of *Adapis parisiensis* (Gingerich & Martin, 1981). Given the absence of this feature in other adapids and omomyids and its somewhat variable distribution among modern lemurs, lorises, galagos and tarsiers (Cartmill, 1978), an exposed orbital ethmoid does not represent a very efficient diagnostic character for Euprimates.

Recently, MacPhee *et al.* (1983) have raised doubts about identifications in fossil forms of the one wholly unique euprimate character, a petrosal auditory bulla (Figure 3). These

Table 1 **Derived features diagnostic for Euprimates**

Features universally present for extant euprimates and unique for mammals
- — osseous auditory bulla completed by an outgrowth from the petrosal (MacPhee & Cartmill, 1986)

Features universally present for extant euprimates and unusual for mammals
- — intratympanic portions of the facial nerve and carotid circulation enclosed within canals by outgrowths of the petrosal (MacPhee, 1981)
- — convergent orbits
- — postorbital bar
- — flattened nail on the hallux

Features nearly universally present for extant euprimate groups and unusual for mammals
- — exposure of ethmoid (os planum) in the orbit (Cartmill, 1978)
- — distinct lateral trochlear margin of the humerus (Szalay & Dagosto, 1980)
- — opposable hallux
- — anterior (distal) elongation of the calcaneum (Martin, 1979)
- — well-developed groove for the tendon of flexor digitorum fibularis on the plantar surface of the calcaneal sustentacular process (Szalay & Drawhorn, 1980)

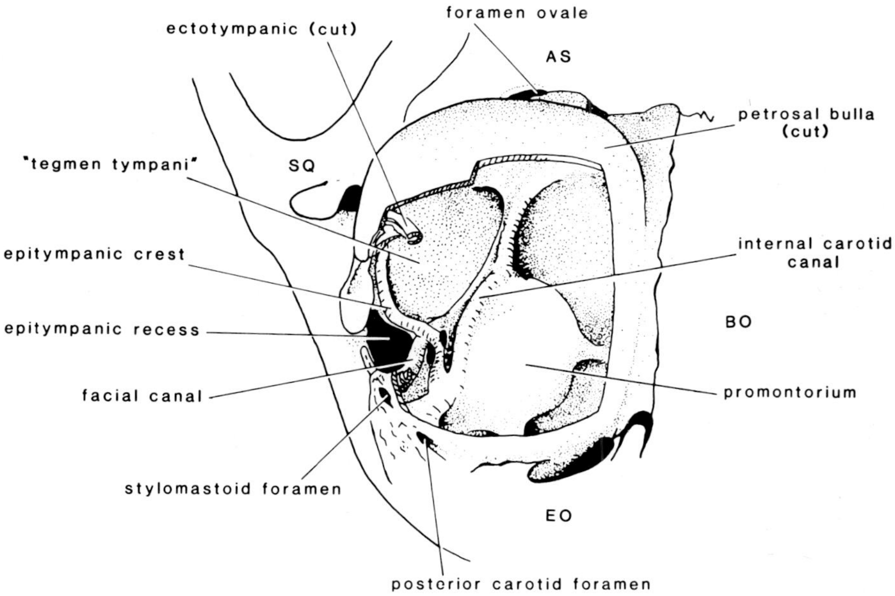

Figure 3. Diagrammatic representation of basicranial structures in an adapid euprimate (modified after MacPhee & Cartmill, 1986). The petrosal bulla and ectotympanic are cut and the middle-ear ossicles removed to expose the hard-tissue contents of the middle-ear space. Not labelled is the stapedial artery, which arises from the internal carotid on the posterior aspect of the promontorium of the petrosal and runs anterolaterally within a canal (a portion of the stapedial canal passes through the epitympanic crest). The portion of the tympanic roof immediately in front of the epitympanic crest is derived from the tegmen tympani, an element of the developing chondrocranium (and not properly an element of the adult skull). Abbreviations: AS—alisphenoid; BO—basioccipital; EO—exoccipital; SQ—squamosal.

authors specifically address this problem in certain archaic primates (see below), but there is a general issue involved here. They argue that it is only through analyses of prenatal development that an osseous auditory bulla can be properly established as a petrosal outgrowth. "It is not possible to infer anything about bullar composition from the absence of bullar sutures in adults; a number of extant mammals (for example, some rodents) develop a non-petrosal bulla which fuses seamlessly with the petrosal in postnatal life" (MacPhee *et al.*, 1983, p. 510). Their point is valid; Klaauw (1931) cites numerous examples among modern mammals where an auditory bulla composed of either an expanded tympanic ring (the ectotympanic bone) or an independent entotympanic bone fuses with the surface of the petrosal. In fact, such fusion is not a rare event; among rodents, complete fusion of the expanded ectotympanic and petrosal is characteristic for nearly half the modern superfamilies (Lavocat & Parent, 1985). Nonetheless, we still hold, as do MacPhee *et al.* (1983), that the auditory bulla in adapids and omomyids is a petrosal outgrowth because it is the most parsimonious solution given current concepts of euprimate interrelationships (Figure 2). If adapids and omomyids have a petrosal bulla, then only a single event (i.e., the appearance of a petrosal bulla in basal euprimates) is required to account for the observed distribution. On the other hand, if adapids and omomyids have an ectotympanic or entotympanic bulla secondarily fused with the petrosal, then multiple events are required (e.g., the appearance of a petrosal bulla in basal euprimates and the loss of that element independently in adapids and omomyids).

Plesiadapiformes

The plesiadapiforms, the archaic primates, include a diverse array of Late Cretaceous to Late Eocene fossil forms, most of which are known exclusively from jaws and teeth. Whole or partial crania are known for only five genera, and although numerous postcrania have been assigned to archaic primates, postcrania have been found in association with dentitions for only one genus, *Plesiadapis* (Szalay & Delson, 1979; Gunnell & Gingerich, 1987). Two major questions concerning the archaic primates must be addressed: (1) are they monophyletic and (2) are some or all of them the closest relatives of the euprimates?

Monophyly. Firstly, there is disagreement about what to group in the archaic primates. Some authors (Szalay & Delson, 1979) include paromomyids, picrodontids, plesiadapids, carpolestids, and saxonellids; others (Gingerich, 1986) add microsyopids (and possibly apatemyids). Because most forms are known almost exclusively from jaws and teeth, effective diagnoses for archaic primates necessarily are restricted to dental features. A perusal of the literature reveals that such diagnoses are wanting. Instead of distinct, identifiable shared derived features, diagnoses of archaic primate dentitions consist of trends, such as reduction of stylar shelf, lower, more bulbous cusps, reduction of trigonid height (Szalay, 1969). Such trends are not unique to archaic primates, but occur in other groups of Paleogene eutherians, for instance, paramyid rodents (Wood, 1962). Perhaps the dentitions of archaic primates cannot be distinguished by discrete, identifiable shared derived features from those of the various Paleogene groups that over the years have been thought to be closely related to archaic primates. If so, then Plesiadapiformes, with or without microsyopids (or apatemyids), is not a monophyletic taxon.

Relationships. Are some archaic primates the closest relatives of euprimates? Evidence of three sorts has been used in the past to ally archaic primates and euprimates: (1) dental, (2) basicranial, and (3) postcranial.

(1) Dental evidence: as a group, the archaic primates have been linked to euprimates on the basis of the sorts of dental trends alluded to above. In addition, as has been noted by numerous authors (Gidley, 1923; Simpson, 1935; Gingerich, 1986), there are very striking similarities in the molars of some advanced plesiadapiforms and primitive euprimates, in particular, the plesiadapid *Plesiadapis* and the notharctine adapid *Cantius*. Gingerich (1986) argues that the resemblances between these forms are so minute that it is difficult to interpret them as homoplasies. Such statements have specific implications for hypotheses of phylogenetic relationship. For one, because many of the derived molar features shared by *Plesiadapis* and *Cantius* are lacking in other plesiadapiforms, one interpretation is that *Plesiadapis* or possibly Plesiadapidae is the sister group to notharctines [Figure 4(a)]. (Any phylogenetic link between *Plesiadapis* and *Cantius* must be through some unknown primitive plesiadapid, because the most primitive known plesiadapid lacks the upper and lower first premolars retained by *Cantius*.) Most, if not all, researchers would reject this hypothesis, because notharctines share a number of derived cranial and skeletal features with adapine and omomyid euprimates absent in plesiadapids.

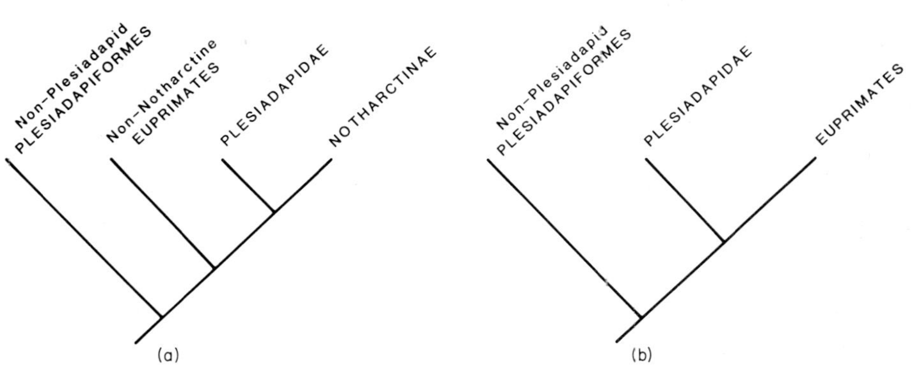

Figure 4. Implications for phylogenetic relationships accepting that the molar similarities of the plesiadapid *Plesiadapis* and the notharctine *Cantius* are synapomorphies. Either: (a) Plesiadapidae and Notharctinae are sister taxa; or (b) Plesiadapidae is the sister group to Euprimates (see text).

A second interpretation of these molar similarities is that *Plesiadapis* or Plesiadapidae is the sister group of Euprimates [Figure 4(b)]. Under this hypothesis, all dental similarities between *Plesiadapis* and *Cantius* are retained primitive euprimate features and necessarily were present in the common ancestor of omomyid, notharctine, and adapine euprimates. This is not the proper forum for a detailed discussion of the dental morphologies of the most primitive omomyid (*Teilhardina*), adapine (*Donrussellia*), and notharctine (*Cantius*), but until a detailed character analysis of the molars of these forms is completed, we believe it is premature to accept *Cantius* as more primitive than the other forms. More important, recent analyses by Covert (1985) reveal that more primitive species of *Cantius* (*C. eppsi*, *C. ralstoni*, and *C. torresi*) do not exhibit the derived molar features shared by *Plesiadapis* and more derived species of *Cantius*. These similarities must have been independently acquired.

(2) Basicranial evidence: basicrania have been described in detail for the paromomyids *Phenacolemur jepseni* (Szalay, 1972) and *Ignacius graybullianus* (MacPhee *et al.*, 1983), the

plesiadapid *Plesiadapis* (Saban, 1963; Russell, 1964; MacPhee *et al.*, 1983), and the microsyopid *Microsyops latidens* (Szalay, 1969). It is generally (Szalay, 1975; MacPhee & Cartmill, 1986) agreed that there are two basic types of basicrania among these archaic primates: a microsyopid type, which is very close to that of ancestral eutherians, and a more derived paromomyid-plesiadapid type. It is this observation that has provided the most substantial support for excluding microsyopids from Plesiadapiformes (Szalay, 1975).

Do either of these two archaic primate basicranial types share unique resemblances with those of euprimates? Two basicranial features are included among the euprimate characters listed in Table 1: a petrosal bulla and canals around the intratympanic portions of the carotid circulation and facial nerve. Microsyopids lack both features. An osseous auditory bulla is not known for any microsyopid (Szalay, 1969). *Microsyops latidens* does have a slight ridge along the medial aspect of the promontorium of the petrosal (Szalay, 1969) that corresponds in position to the area where the petrosal plate develops in modern euprimates (MacPhee, 1981; MacPhee & Cartmill, 1986). Yet, this ridge clearly is not a petrosal bulla nor is it uncommon in basicrania of other Paleogene eutherians (Novacek, 1986; MacPhee *et al.*, in press). Also lacking in *M. latidens* are canals for the intratympanic portions of the carotid circulation and facial nerve; these structures ran in open sulci on the ventral surface of the petrosal (Szalay, 1969). One of us (J.R.W.) has recently had the opportunity to study a beautifully preserved skull of the microsyopid *Microsyops annectens* from the Middle Eocene of Wyoming (Geological Museum, University of Wyoming No. 12363) collected by J. G. Eaton (a full description of this specimen by M. J. Novacek and J. G. Eaton is in preparation). The basicranium of this specimen exhibits the same suite of characters described above for *M. latidens*.

Identification of the two euprimate basicranial characters in known paromomyids and plesiadapids is more problematic. Closed canals for portions of the carotid circulation have been described (Russell, 1964; Szalay, 1972, 1975), but more recently, MacPhee *et al.* (1983) have argued that these "canals" are imperforate. Having not studied all specimens, a choice between these two views is not possible, but we agree with MacPhee & Cartmill (1986) that the size of the possible routes for the internal carotid and stapedial arteries indicate that these vessels were probably vestigial in adult paromomyids and plesiadapids. Although the identification of arterial canals is questionable, there is clearly no canal for the intratympanic portion of the facial nerve in either paromomyids (Szalay, 1972) or plesiadapids (Saban, 1963). Evidence for a complete osseous auditory bulla continuous with the petrosal is apparent in basicrania of both paromomyids and plesiadapids. However, the previously cited caution of MacPhee *et al.* (1983) concerning the identification of petrosal bullae in fossils is still relevant here. Unlike the adapid-omomyid example discussed above, parsimony does not resolve this problem for the condition of the paromomyid-plesiadapid bulla. An equal number of steps is required to obtain either a petrosal bulla or a non-petrosal bulla fused secondarily to the petrosal in paromomyids and plesiadapids from ancestral conditions of an entotympanic, ectotympanic, or no osseous bulla (Figure 5). We agree with MacPhee *et al.* that the presence of a petrosal bulla is not demonstrable in paromomyids and plesiadapids.

The only other region of the cranium that supplies characters allying some archaic primates to euprimates is the wall of the orbit. In most eutherians and in marsupials, the maxilla makes no contribution to the orbital wall; it is excluded by the large orbital process of the palatine contacting the frontal and lacrimal (Novacek, 1986). The orbits of the

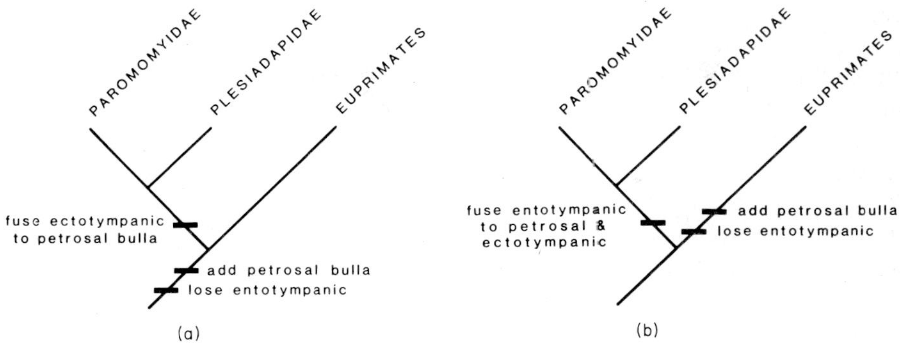

Figure 5. Character phylogenies accounting for the bullar morphology found in paromomyids, plesiadapids, and euprimates. In the examples shown here, an entotympanic bulla is taken as the primitive state, but either an ectotympanic bulla or no osseous bulla as the primitive state yields similar results. The bullae of paromomyids and plesiadapids have no visible sutures with either the promontorium of the petrosal or ectotympanic bone. [Unlike the separate, ring-shaped element of lemuriforms and adapids (Figure 1), the ectotympanic in paromomyids and plesiadapids is part of the bulla, as in lorisiforms, omomyids, tarsiers, and anthropoids.] In the examples here, the paromomyid-plesiadapid bulla might be either (a) a derived petrosal bulla secondarily fused to the ectotympanic or (b) a retained entotympanic bulla secondarily fused to the promontorium of the petrosal and ectotympanic. In the former, a petrosal bulla is a synapomorphy of paromomyids, plesiadapids, and euprimates; in the latter, the petrosal bulla is restricted to euprimates. An equal number of steps is required to account for either.

microsyopids *Microsyops lundeliusi* (Szalay, 1969) and *M. annectans* (University of Wyoming No. 12363) exhibit this pattern. In contrast, the plesiadapid *Plesiadapis*, the only other archaic primate whose orbit is known in detail, has a large orbital process of the maxilla that contacts the frontal; the palatine is confined to the floor of the orbit, although it does contact the frontal (Russell, 1964; Gingerich, 1976). The same pattern occurs in adapids (Le Gros Clark, 1934; Russell, 1964), omomyids (Simons & Russell, 1960; Wilson, 1966), and some modern euprimates (Cartmill, 1978). (An orbital process of the maxilla occurs in most modern euprimates, but usually does not contact the frontal because of the intervening orbital exposure of the ethmoid.) Although a maxillary orbital process contacting the frontal is a shared derived feature of *Plesiadapis* and euprimates, it is not a very unusual one among eutherians, appearing in lipotyphlan insectivorans, rodents, lagomorphs, hyracoids, proboscideans, and some edentates (Novacek, 1985*b*, 1986; Wible, unpublished data).

(3) Postcranial evidence. The tarsals of *Plesiadapis* (*P. tricuspidens* and *P. gidleyi*), specifically the astragalus and calcaneum, exhibit a number of specializations also found in euprimates that are associated with free mobility of the foot (Szalay, 1977; Szalay & Drawhorn, 1980). Most of these specializations, however, are not unique to *Plesiadapis* and euprimates, but are also known for modern scandentians (tree shrews) and dermopterans (colugos or flying lemurs). The only derived tarsal feature, according to Szalay & Drawhorn (1980), that distinguishes *Plesiadapis* and euprimates from tree shrews and dermopterans is the presence of a more pronounced groove for the tendon of flexor digitorum fibularis on the plantar surface of the calcaneum. We have doubts about the reliability of this feature. First of all, although relatively unusual among mammals, a well-developed groove and presumed concomitant increased muscle mass for flexor

fibularis also appear, for instance, in some lagomorphs and rodents (Covert, unpublished data). Second and more importantly, a well-developed groove for flexor fibularis does not appear in the calcanea of all *Plesiadapis*. One of us (J.R.W.) has recently had the opportunity to study the skull and associated partial skeleton of *Plesiadapis cookei* from the Early Eocene of Wyoming (University of Michigan Nos. 87990 and 69934), which was discovered in 1986 and reported in Gunnell & Gingerich (1987) (a full description of this specimen by these authors is in preparation). The right calcaneum, the only one preserved, does not show any pronounced excavation for the tendon of flexor fibularis. Given this variability, a well-developed groove for flexor fibularis is not a reliable character for allying *Plesiadapis* and euprimates.

In summary, the evidence for allying the various archaic primates and euprimates consists of shared dental trends, none of which is unique to these groups among Eutheria. One archaic primate, the plesiadapid *Plesiadapis*, provides additional evidence in that it shares several derived features of the orbit and tarsus with euprimates, but again these are not unique to these forms. Among the archaic primates, paromomyids and plesiadapids are distinguished from microsyopids by several shared derived features of the basicranium (e.g., reduction of the internal carotid and stapedial arteries, a bulla continuous with the petrosal). Derived basicranial features also distinguish euprimates from microsyopids (e.g., canals around the internal carotid and stapedial arteries, a petrosal bulla). However, the evidence from the basicranium is equivocal with regard to the affinities of euprimates with paromomyids and plesadapids.

Superordinal relationships

In the past twenty years, morphological and molecular studies have provided support for allying primates with a diverse array of modern eutherian orders and extinct groups. In 1986 alone, primates have been linked with rodents and lagomorphs based on protein sequence data (Miyamoto & Goodman, 1986), with chiropterans, dermopterans, and scandentians based on postcranial features (Novacek & Wyss, 1986), with megachiropterans (Old World fruit bats) and dermopterans based on neuroanatomical data (Pettigrew & Cooper, 1986), and with scandentians based on immunodiffusion distance data (Shoshani, 1986). For the molecular and neuroanatomical studies, the grouping primates is equivalent to the monophyletic taxon Euprimates. However, unless explicitly stated, it is not clear what the grouping primates includes to the paleontologist; although their analysis suggests that some plesiadapiforms may be included, Novacek & Wyss (1986), for instance, do not explicitly define their usage. Several current hypotheses of the higher-level relationships of primates (*sensu lato*) are reviewed in the following.

Rodents and lagomorphs. Similarities in the molars of Paleocene paramyid rodents and plesiadapiform primates led McKenna (1961) and Wood (1962) to propose an origin for rodents within primates (*sensu lato*). McKenna (1975) later abandoned this hypothesis, and support for a primate-rodent clade from other morphological systems has not appeared. The most widely held current view based on data from various morphological systems is that rodents should be grouped with lagomorphs in the Superorder Glires (Luckett & Hartenberger, 1985). Additionally, several researchers (Novacek, 1982, 1986; Novacek & Wyss, 1986; Shoshani, 1986; Wible, in press) have offered considerable support from cranial and postcranial anatomy for Macroscelidea (elephant shrews) as the sister group to Glires.

Molecular studies provide diverse views of the relationships of rodents and lagomorphs. Albumin immunological distance data provide no support for Glires, but ally rodents with bats, carnivorans, and probably talpids (moles) and soricids (shrews) (Sarich, 1985). Sequence analysis of the α-crystallin A eye-lens protein does not contradict Glires, but places the closest relatives of rodents among the lagomorphs, euprimates, lipotyphlan insectivorans, and tree shrews (De Jong, 1985). Sequence analysis of alpha and beta hemoglobins combined in tandem alignment supports a grouping of lagomorphs, bats, and tree shrews (Shoshani *et al.*, 1985), while analysis of seven proteins combined in tandem alignment advocates a trichotomy for rodents, lagomorphs, and euprimates (Miyamoto & Goodman, 1986).

Of the features discussed in our review, *Plesiadapis* and euprimates share with rodents and lagomorphs a derived orbital process of the maxilla. Although this apomorphy supports the rodent-lagomorph-euprimate clade proposed from tandem alignment analysis by Miyamoto & Goodman (1986), we do not find this phylogenetic hypothesis compelling. A maxillary orbital process is not unique to these forms, but appears in several other eutherian groups. Also, several authors (Maeda & Fitch, 1981; Wyss *et al.*, 1987) in recent analyses of tandem alignment sequence data have raised questions about the resolving power of this technique.

Plesiadapis shares an additional derived feature with rodents and lagomorphs; its procumbent upper incisors lie in an enlarged premaxilla with an expanded posterodorsal process contacting the frontal (Russell, 1964; Gingerich, 1976; Novacek, 1985*b*). In other archaic primates (i.e., the microsyopid *Microsyops latidens*, Szalay, 1969; the carpolestid *Carpolestes dubius*, Rose, 1975; the paromomyids *Phenacolemur pagei*, Gingerich, 1976, and *Palaecthon nacimienti*, Kay & Cartmill, 1977), the premaxilla is restricted and does not contact the frontal. One possible exception is the paromomyid *Ignacius graybullianus*, in which the premaxilla may have contacted the frontal, but this region is obscured by fractures (Rose & Gingerich, 1976). Given that an enlarged premaxilla with a posterodorsal process appears in other mammals with procumbent incisors (e.g., the marsupial *Dactylopsila*, the euprimate *Daubentonia*), this feature has presumably been acquired independently on several occasions (Gingerich, 1971).

Erinaceomorph insectivorans. Basicranial similarities have been used to support close affinities for erinaceomorph insectivorans and primates (euprimates and non-microsyopid plesiadapiforms) by Szalay (1975). Particular significance is placed on the large petrosal contribution to the auditory bullae in both groups: most fossil and recent erinaceomorphs have a petrosal bullar component in addition to a large tympanic process of the basisphenoid (Butler, 1948; Rich & Rich, 1971). Novacek (1982) tentatively supported this hypothesis with other shared derived features of the cranium (e.g., a process of the maxilla in the orbital wall contacting the frontal) and with dental resemblances between the earliest known archaic primates, such as the paromomyid *Purgatorius*, and erinaceomorphs.

Both Szalay (1977) and Novacek (Novacek *et al.*, 1983) have abandoned the hypothesis of close primate-erinaceomorph ties. However, Szalay continues to argue that "primitive primates share some significant (basicranial) similarities with the erinaceomorph morphotype" and that it is "likely that primates and other archontans originated from . . . an adapisoricid insectivoran, with a basicranium not very different from that of the erinaceomorph morphotype" (Szalay & Delson, 1979, pp. 26–27). The presumed derived features shared by the primate (*sensu* Szalay, 1975) and erinaceomorph morphotypes are, in addition to the petrosal bullar contribution, a rounded promontorium, bony canals for the entire intrabullar carotid circulation, fenestra rotunda ventrally "shielded" by the internal carotid canal, and loss of the medial internal carotid artery (Szalay & Delson, 1979). The analyses of Presley (1979) and Wible (1984, 1986) have shown that two separate internal carotid arteries (medial and lateral) were not present in ancestral eutherians and, therefore, the loss of one of the two cannot be a derived feature. [Further, the lateral or promontorial position of the single internal carotid artery of modern erinaceomorphs and strepsirhine euprimates is likely a retained plesiomorphous state (Wible, 1986).] A rounded promontorium of the petrosal is of questionable value, as Szalay & Delson rightly point out, because it appears in many mammalian groups. Finally, more recent analyses of the basicrania of non-microsyopid plesiadapiforms and extinct erinaceomorphs reveal that the other features cited by Szalay & Delson are not shared by the primate (*sensu* Szalay, 1975) and erinaceomorph morphotypes. First of all, the Eocene erinaceomorphs *Diacodon* and *Pholidocercus* have insignificant petrosal contributions to the auditory bulla (Novacek *et al.*, 1983; MacPhee & Storch, 1987; MacPhee *et al.*, in press), suggesting that these elements may have been acquired independently in other forms and are not part of the erinaceomorph morphotype. Secondly, bony canals for the intrabullar carotid circulation are absent in paromomyids and plesiadapids (MacPhee *et al.*, 1983), *Diacodon* (Novacek *et al.*, 1983), and *Pholidocercus* (MacPhee & Storch, 1987; MacPhee *et al.*, in press) and are not part of either the primate (*sensu* Szalay, 1975) or erinaceomorph morphotype.

Archonta. Gregory's (1910) Superorder Archonta was revived by McKenna (1975) to include bats, dermopterans, tree shrews, and primates (euprimates and non-microsyopid plesiadapiforms). This grouping has been supported by shared derived features of the tarsus in *Plesiadapis*, euprimates, tree shrews, and dermopterans (Szalay, 1977; Szalay & Drawhorn, 1980) and by albumin and transferrin immunological data linking euprimates to tree shrews and dermopterans (Cronin & Sarich, 1980). Critique of the Archonta has centered on the lack of evidence for the inclusion of bats (Novacek, 1980, 1982; Cartmill & MacPhee, 1980). Recently, however, Novacek & Wyss (1986) have restudied the tarsal bones of bats and have found a derived feature in the megachiropteran astragalus described previously for only dermopterans, tree shrews, *Plesiadapis*, and euprimates by Szalay (1977; Szalay & Drawhorn, 1980). This feature, a distinct medial confluence of the sustentacular and distal astragalar facets, does not appear in any other eutherian, including microchiropterans (echolocating bats), which lose the sustentacular facet altogether. Novacek & Wyss (1986) tentatively support Archonta with this feature and an additional synapomorphy of the male external genitalia. In most recent eutherians, the penis is enclosed in a sheath-like genital pouch that forms a sling in close attachment to the abdomen, but in recent archontans the penis is free and pendulous. Novacek & Wyss caution that although this feature is unique to recent Archonta, conditions vary considerably within members of the group, including some microchiropterans where the penis presumably reverts to the more generalized sheathed condition.

The supporting evidence is not overwhelming, but Archonta currently represents the only hypothesis for the higher-level relationships of primates (euprimates and *Plesiadapis*) based on uniquely derived morphological features. Accepting this hypothesis, can relationships within the superorder be refined further? Figure 6 shows two previously published cladograms of archontan interrelationships. The two differ radically in that

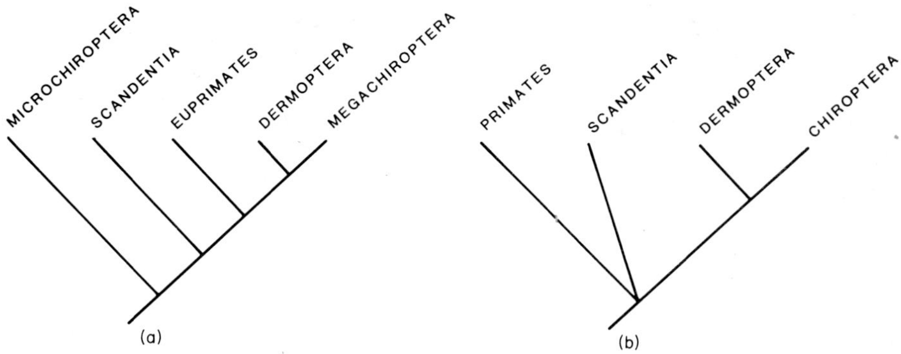

Figure 6. Previously published hypotheses of archontan interrelationships. (a) From Smith & Madkour (1980), based chiefly on features of the penis. (b) From Novacek & Wyss (1986), based on cranial and postcranial features. Neither set of authors specifically discusses the relationships of the archaic primates, although the distribution of characters discussed by Novacek & Wyss suggests that some plesiadapiforms may be included within primates.

Figure 6(a) depicts Chiroptera as diphyletic, while in Figure 6(b) bats are monophyletic. Bat diphyly, as proposed by Smith & Madkour (1980), Pettigrew (1986), and Pettigrew & Cooper (1986), is supported principally by differences in the flight apparatuses of mega- and microchiropterans and by derived features that megachiropterans share with euprimates and dermopterans that are absent in microchiropterans. Differences in the wings of mega- and microchiropterans have been recorded (Strickler, 1978), but are not unexpected, given the more than 50 million years of independent evolution of these groups within the Order Chiroptera (Novacek, 1985*a*). Moreover, these slight differences are overshadowed by the many unique cranial and postcranial features that are restricted to mega- and microchiropterans among eutherians (Van Valen, 1979; Novacek, 1982; Wible, 1984, in press; Wible & Novacek, in press) and by supporting albumin immunological data (Cronin & Sarich, 1980). The derived features of megachiropterans, euprimates, and dermopterans identified by Smith & Madkour (1980), Pettigrew (1986), and Pettigrew & Cooper (1986) alternatively may be plesiomorphous for Archonta or some subset thereof and lost secondarily in microchiropterans (see below).

The only refinement of relationships within Archonta offered by Novacek & Wyss (1986) is an alliance of Chiroptera and Dermoptera [Figure 6(b)]. This relationship seems secure, with support from features of both cranial and postcranial anatomy (see also Novacek, 1986). A Chiroptera-Dermoptera clade is accepted for the following evaluation of several alternative hypotheses of archontan interrelationships.

(1) A grouping of euprimates and tree shrews [Figure 7(a)] is supported by several shared derived features of the cranium. Included are two euprimate synapomorphies from Table 1: a postorbital bar and canals around the intratympanic portions of the facial nerve and stapedial artery formed from outgrowths of the petrosal (the canal around the internal carotid artery is completed by an entotympanic element in tree shrews) (MacPhee, 1981). Other derived features shared by euprimates and tree shrews include an anterior carotid foramen in the basisphenoid that is converted into a long tube and a tegmen tympani that is greatly expanded anterolaterally to cover the entire middle-ear ossicular chain (MacPhee, 1981) (Figure 3). The anterior carotid foramen, the opening through which the internal carotid artery enters the braincase, is usually a simple opening either within the

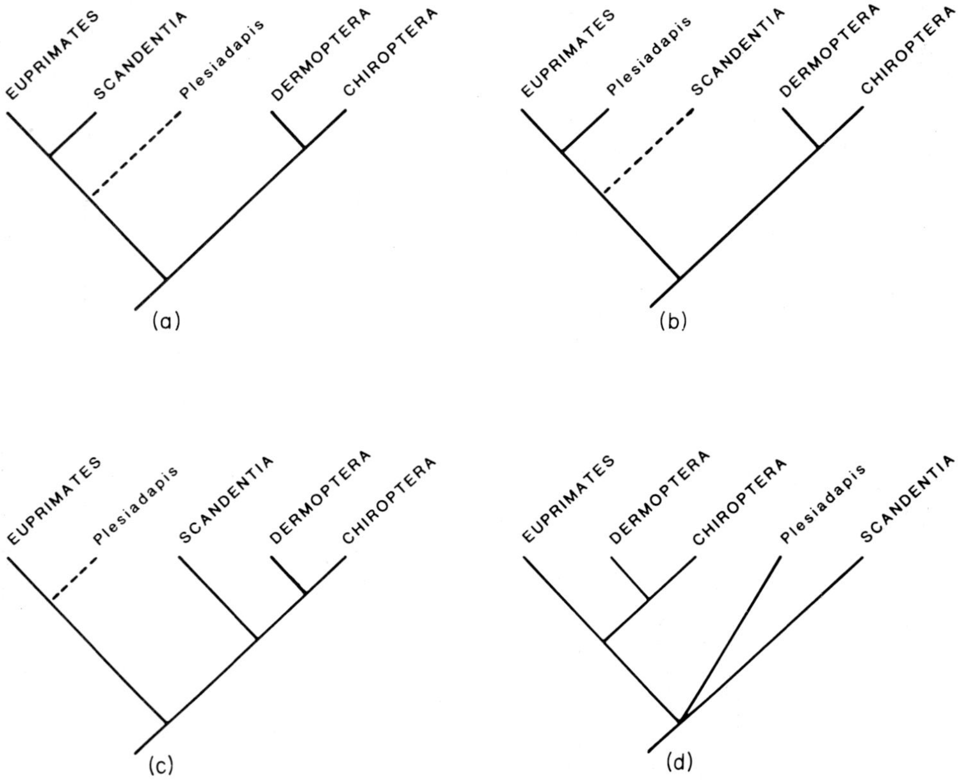

Figure 7. Additional hypotheses of archontan interrelationships, all of which have some morphological basis (see text). Dashed line denotes uncertainty of relationships.

basisphenoid or between the basisphenoid and petrosal. A tubular anterior carotid foramen appears to be somewhat unusual among other eutherians, although it occurs in brachyericine erinaceomorphs (Rich & Rich, 1971) and some microchiropterans (Wible, 1984). The tegmen tympani, an element of the mammalian chondrocranium projecting forward from the anterolateral portion of the auditory capsule to form the posterior portion of the tympanic roof, is typically small in eutherians (De Beer, 1937), but not in tree shrews and euprimates.[1] In addition to its large size, the tegmen tympani in modern tree shrews and strepsirhine primates is further characterized by the presence of an epitympanic crest, an incomplete osseous septum bordering the anterior margin of the epitympanic recess within which the stapedial artery runs (MacPhee, 1981; Zeller, 1986) (Figure 3). Among other archontans, the tegmen tympani is lost in dermopterans (Halbsguth, 1973) and is a

[1] Elephant shrews (MacPhee, 1981), lagomorphs (Wible, 1984), and rodents, including Early Tertiary forms (Wahlert, 1974; Parent, 1983; Wible, unpublished data) are the only other eutherians with a tegmen tympani expanded to a degree similar to that of tree shrews and primates. However, the expanded tegmen tympani of elephant shrews, lagomorphs, and rodents differs from that of tree shrews and primates with regard to the course of the lesser petrosal nerve and ramus inferior of the stapedial artery and so, presumably has been acquired independently. In the former the lesser petrosal nerve and ramus inferior run dorsal to the tegmen tympani, a derived pattern within Eutheria, while in the latter they are ventral (Wible, 1984, in press). A superordinal grouping of elephant shrews, lagomorphs, and rodents is supported by this and numerous other cranial and postcranial features (Novacek, 1982, 1986; Novacek & Wyss, 1986).

slender, vertically-oriented rod in chiropterans, a bat synapomorphy (Wible, 1984, in press; Wible & Novacek, in press). A tubular anterior carotid foramen and expanded tegmen tympani have not been identified for *Plesiadapis* or other archaic primates. The tympanic roof is relatively broad in *Plesiadapis* (Saban, 1963; Russell, 1964) and *Phenacolemur* (Szalay, 1972), but sutures defining the limits of the petrosal contribution are not evident.

(2) To ally euprimates and *Plesiadapis* [Figure 7(b)], the shared orbital process of the maxilla must be weighted more heavily than the euprimate-tree shrew synapomorphies discussed above. Given that this feature is not unique to euprimates and *Plesiadapis*, we are not convinced that a cladogram grouping these forms is more resolved than one grouping euprimates and tree shrews.

(3) The only presumed derived feature identified for tree shrews, dermopterans, and bats [Figure 7(c)] is an auditory bulla formed from independent entotympanic elements (Novacek, 1977; Cartmill & MacPhee, 1980). Yet, the homologies of the entotympanics in these groups are uncertain; developmentally, tree shrews have only one entotympanic element (MacPhee, 1981; Zeller, 1986), dermopterans and microchiropterans have two (Wible, 1984), and megachiropterans have up to four (Klaauw, 1922; Wible, 1984). As an alternative, given the presence of entotympanics in several archontan outgroups (e.g., Xenarthra, Carnivora, Leptictida), these elements merely may be primitive for Archonta.

(4) The grouping of euprimates with the dermopteran-chiropteran clade [Figure 7(d)] is supported chiefly by derived penile and neuroanatomical characters of euprimates, dermopterans, and megachiropterans (Smith & Madkour, 1980; Pettigrew, 1986; Pettigrew & Cooper, 1986). However, to maintain bat monophyly, these features must be lost secondarily in microchiropterans. We suggest caution here, because, in particular, the distribution of these characters is known for a rather limited number of forms and orders. The penile character concerns the corpus spongiosum, the erectile tissue that surrounds the urethra. In the state held to be primitive for Eutheria by Smith & Madkour (1980), the corpus spongiosum is not well organized and is restricted to the proximal portion of the penis. In contrast, in euprimates, dermopterans, and megachiropterans, the corpus spongiosum is highly vascularized and expanded into the glans of the penis. Yet, the homologies of this feature seem questionable, given that the degree of expansion is somewhat variable. Moreover, in most microchiropterans studied by Smith & Madkour, the corpus spongiosum is present in the glans to some extent, just not expanded. The neuroanatomical characters concern the projections of various pathways in the visual system (Pettigrew, 1986; Pettigrew & Cooper, 1986). Detailed descriptions of these characters should be completed for other eutherians before assessing phylogenetic significance.

Where do microsyopids fit? Szalay & Drawhorn (1980) have allied these forms to the Archonta. In fact, these authors speculate that microsyopids may have dermopteran affinities. The basis for this speculation is the occurrence of dermopteran-like tarsals in Early Eocene localities for which microsyopid dentitions are known, but for which plagiomenid dentitions are not known [plagiomenids are held to be fossil dermopterans (Rose & Simons, 1977)]. The newly discovered skull of the Middle Eocene *Microsyops annectens* (University of Wyoming 12363) discussed above reveals details of the basicranium that argue against dermopteran affinities. In particular, the roof of the tympanic cavity has a very large contribution from the petrosal; modern dermopterans have virtually no petrosal contribution because the tegmen tympani is absent. Although the mastoid portion of the auditory capsule has a broad, flat eminence on its posterolateral aspect (see Szalay, 1969, Figure 17) resembling the arrangement in recent dermopterans, these processes are not equivalent. In *Cynocephalus* this process is highly pneumatized (Hunt & Korth, 1980) and has canals for a major cranial artery, the arteria diploetica magna, and veins (Wible, unpublished data). Evidence of pnematization is not apparent in *M. annectans*, nor are there any vascular canals. In fact, the arteria diploetica magna, a derived eutherian character (Wible, 1984, in press), appears to be absent in this form, because its usual foramen of entry into the braincase between the squamosal and mastoid is lacking.

The affinities of microsyopids remain muddled. The large petrosal contribution to the tympanic roof is reminiscent of the expanded tegmen tympani in tree shrews and euprimates, yet an epitympanic crest is lacking. Other portions of the basicranium and orbit come close to the pattern reconstructed for primitive eutherians by Novacek (1986). Until postcrania found in association with microsyopid dentitions confirm the speculations of Szalay & Drawhorn (1980), we see no basis from the cranium for linking these forms with either Dermoptera or Archonta.

Conclusions

The following conclusions derive from the foregoing cladistic analysis:

(1) Euprimates is a monophyletic group diagnosed by a suite of shared derived characters of the cranium and postcranium exhibiting unique or relatively unique distributions among Eutheria.

(2) Plesiadapiformes (with or without Microsyopidae) is not demonstrably a monophyletic group. Paromomyidae and Plesiadapidae form a monophyletic group within Plesiadapiformes diagnosed by several shared derived characters of the basicranium.

(3) The Superorder Archonta (Euprimates, Scandentia, Dermoptera, and Chiroptera) is the only higher-level grouping including Euprimates that is diagnosed by unique shared derived morphological characters. The plesiadapid *Plesiadapis*, the only plesiadapiform for which postcrania have been found with dentitions, exhibits the unique tarsal character that unites Archonta, supporting the inclusion of Plesiadapidae (and Paromomyidae) within Archonta. Microsyopidae does not demonstrably have archontan affinities.

(4) Hypotheses of relationships within Archonta ally Euprimates with either Scandentia or *Plesiadapis* (Paromomyidae-Plesiadapidae). The Euprimates-Scandentia clade is supported by shared derived characters of the cranium, several of which appear to be unique among Eutheria. The Euprimates-*Plesiadapis* clade is supported by a shared derived character of the orbit, which is also found in a number of other eutherians. Relying on derived characters that exhibit little homoplasy among other eutherian orders, the Euprimates-Scandentia hypothesis is preferable.

(5) Following the criterion for restricting higher taxonomic groups to the crown group, the Order Primates is equivalent to Euprimates. The euprimate stem lineage is not demonstrable; it may include *Plesiadapis* (Paromomyidae-Plesiadapidae), but this is equivocal. Therefore, the criterion of equating higher taxonomic groups with the crown group and its stem lineage cannot be meaningfully employed.

Acknowledgements

The authors thank T. M. Bown (U.S. Geological Survey, Denver Station), P. D. Gingerich (University of Michigan, Ann Arbor), and M. J. Novacek (American Museum of Natural

History, New York) for access to specimens in their care. For helpful discussions and comments on various portions of the manuscript, we are grateful to M. Cartmill, J. M. Clark, J. A. Hopson, J. M. Humphreys, R. D. E. MacPhee, M. J. Novacek, L. Van Valen, R. T. Zanon, and four anonymous reviewers. The senior author acknowledges support from National Science Foundation grant DEB-8208797.

References

Ax, P. (1985). Stem species and the stem lineage concept. *Cladistics* **1,** 279–287.

Butler, P. M. (1948). On the evolution of the skull and teeth in the Erinaceidae, with special reference to fossil material in the British Museum. *Proc. zool. Soc. Lond.* 1948, 446–500.

Cartmill, M. (1972). Arboreal adaptations and the origin of the Order Primates. In (R. Tuttle, Ed.) *The Functional and Evolutionary Biology of Primates*, pp. 97–122. Chicago: Aldine-Atherton.

Cartmill, M. (1974). Rethinking primate origins. *Science* **184,** 436–443.

Cartmill, M. (1978). The orbital mosaic in prosimians and the use of variable traits in systematics. *Folia primat.* **30,** 89–114.

Cartmill, M. & MacPhee, R. D. E. (1980). Tupaiid affinities: the evidence of the carotid arteries and cranial skeleton. In (W. P. Luckett, Ed.) *Comparative Biology and Evolutionary Relationships of Tree Shrews*, pp. 95–132. New York: Plenum Press.

Covert, H. H. (1985). Adaptations and evolutionary relationships of the Eocene Primate Family Northarctidae. Ph.D. Dissertation, Duke University.

Cronin, J. E. & Sarich, V. M. (1980). Tupaiid and Archonta phylogeny: the macromolecular evidence. In (W. P. Luckett, Ed.) *Comparative Biology and Evolutionary Relationships of Tree Shrews*, pp. 293–312. New York: Plenum Press.

De Beer, G. R. (1937). *The Development of the Vertebrate Skull.* Oxford: Clarendon Press.

De Jong, W. W. (1985). Superordinal affinities of Rodentia studied by sequence analysis of eye lens protein. In (W. P. Luckett & J.-L. Hartenberger, Eds) *Evolutionary Relationships Among Rodents: A Multidisciplinary Analysis*, pp. 211–226. New York: Plenum Press.

Gauthier, J. (1986). Saurischian monophyly and the origin of birds. In (K. Padian, Ed.) *The Origin of Birds and the Evolution of Flight. Mem. Cal. Acad. Sci.*, **8,** 1–55.

Gauthier, J., Estes, R. & De Queiroz, K. (in press). A phylogenetic analysis of Lepidosauromorpha. In (R. Estes & G. Pregill, Eds) *The Phylogenetic Relationships of the Lizard Families.* Palo Alto: Stanford University Press.

Ghiselin, M. T. (1984). "Definition", "character", and other equivocal terms. *Syst. Zool.* **33,** 104–110.

Gidley, J. W. (1923). Paleocene primates of the Fort Union, with discussion of relationships of Eocene primates. *Proc. US natn. Mus.* **63,** 1–38.

Gingerich, P. D. (1971). Cranium of the Late Palaeocene primate *Plesiadapis tricuspidens. Nature* **230,** 324–325.

Gingerich, P. D. (1976). Cranial anatomy and evolution of early Tertiary Plesiadapidae. *Mus. Paleont. Univ. Mich., Pap. Paleont.* **15,** 1–140.

Gingerich, P. D. (1986). *Plesiadapis* and the delineation of the Order Primates. In (B. A. Wood, L. B. Martin & P. Andrews, Eds) *Major Topics in Primate and Human Evolution,* pp. 32–46. Cambridge: Cambridge University Press.

Gingerich, P. D. & Martin, R. D. (1981). Cranial morphology and adaptations in Eocene Adapidae. II. The Cambridge skull of *Adapis parisiensis. Am. J. phys. Anthrop.* **56,** 235–257.

Goodman, M. (Ed.) (1982). *Macromolecular Sequences in Systematic and Evolutionary Biology.* New York: Plenum Press.

Gregory, W. K. (1910). The orders of mammals. *Bull. Am. Mus. nat. Hist.* **27,** 1–524.

Gunnell, G. F. & Gingerich, P. D. (1987). Skull and partial skeleton of *Plesiadapis cookei* from the Clark's Fork Basin, Wyoming. *Am. J. phys. Anthrop.* **72,** 206.

Halbsguth, A. (1973). Das Cranium eines Foeten des Flattermaki *Cynocephalus volans (Galeopithecus volans)* (Mammalia. Dermoptera) von 63 mm SchStlg. Inaugural-Dissertation (Medizin), Johann Wolfgang Goethe-Universität.

Hennig, W. (1966). *Phylogenetic Systematics.* Urbana: University of Illinois Press.

Hennig, W. (1981). *Insect Phylogeny.* Chichester: John Wiley & Sons.

Hoffstetter, R. (1977). Phylogénie des Primates: confrontation des résultats obtenus par les diverse voies d'approches de problème. *Bull. Mém. Soc. Anthrop. Paris, sér. 13,* **4,** 327–346.

Hunt, R. M., Jr & Korth, W. W. (1980). The auditory region of Dermoptera: morphology and function relative to other living mammals. *J. Morph.* **164,** 167–211.

Kay, R. F. & Cartmill, M. (1977). Cranial morphology and adaptations of *Palaecthon nacimienti* and other Parmomyidae (Plesiadapoidea, ? Primates), with a description of a new genus and species. *J. hum. Evol.* **6,** 19–53.

Klaauw, J. C. van der (1922). Über die Entwicklung des Entotympanicums. *Tijdschr. ned. dierk. Vereen.* **18,** 135–174.

Klaauw, J. C. van der (1931). The auditory bulla in some fossil mammals, with a general introduction to this region of the skull. *Bull. Am. Mus. nat. Hist.* **62,** 1–352.
Lavocat, R. & Parent, J.-P. (1985). Phylogenetic analysis of middle ear features in fossil and living rodents. In (W. P. Luckett & J.-L. Hartenberger, Eds) *Evolutionary Relationships Among Rodents: A Multidisciplinary Analysis,* pp. 333–354. New York: Plenum Press.
Lauder, G. (1981). Form and function: structural analysis in evolutionary morphology. *Paleobiology* **7,** 430–442.
Le Gros Clark, W. E. (1934). On the skull structure of *Pronycticebus gaudryi. Proc. zool. Soc. Lond.* 1934, 19–27.
Le Gros Clark, W. E. (1959). *The Antecedents of Man: An Introduction to the Evolution of the Primates.* Edinburgh: Edinburgh University Press.
Lillegraven, J. A., Kielan-Jaworowska, Z. & Clemens, W. A. (Eds) (1979). *Mesozoic Mammals: The First Two-Thirds of Mammalian History.* Berkeley: University of California Press.
Luckett, W. P. & Hartenberger, J.-L. (1985). Evolutionary relationships among rodents: comments and conclusions. In (W. P. Luckett & J.-L. Hartenberger, Eds) *Evolutionary Relationships Among Rodents: A Multidisciplinary Analysis,* pp. 685–712. New York: Plenum Press.
MacPhee, R. D. E. (1981). Auditory regions of primates and eutherian insectivores: morphology, ontogeny, and character analysis. *Contr. Primat.* **19,** 1–282.
MacPhee, R. D. E. & Cartmill, M. (1986). Basicranial structures and primate systematics. In (D. R. Swindler & J. Erwin, Eds) *Comparative Primate Biology.* Vol. 1, *Systematics, Evolution, and Anatomy,* pp. 219–275. New York: Alan R. Liss, Inc.
MacPhee, R. D. E., Cartmill, M. & Gingerich, P. D. (1983). New Palaeogene primate basicrania and the definition of the Order Primates. *Nature* **301,** 509–511.
MacPhee, R. D. E. & Storch, G. (1987). Primate origins and basicranial morphology: significance of Middle Eocene erinaceomorphs *Pholidocercus* and *Macrocranion. Am. J. phys. Anthrop.* **72,** 229.
MacPhee, R. D. E., Storch, G. & Novacek, M. J. (in press). Basicranial morphology of early Tertiary erinaceomorphs and the origin of primates. *Am. Mus. Novit.*
Maddison, W. P., Donoghue, M. J. & Maddison, D. R. (1984). Outgroup analysis and parsimony. *Syst. Zool.* **33,** 83–103.
Maeda, N. & Fitch, W. M. (1981). Amino acid sequences of a myoglobin from lace monitor lizard, *Varanus varius,* and its evolutionary implications. *J. biol. Chem.* **256,** 4301–4309.
Martin, R. D. (1968). Towards a new definition of primates. *Man* **3,** 377–401.
Martin, R. D. (1979). Phylogenetic aspects of prosimian behavior. In (G. A. Doyle & R. D. Martin, Eds) *The Study of Prosimian Behavior,* pp. 45–77. New York: Academic Press.
Martin, R. D. (1986). Primates: a definition. In (B. A. Wood, L. B. Martin & P. Andrews, Eds) *Major Topics in Primate and Human Evolution,* pp. 1–31. Cambridge: Cambridge University Press.
McKenna, M. C. (1961). A note on the origin of rodents. *Am. Mus. Novit.* **2037,** 1–5.
McKenna, M. C. (1975). Towards a phylogenetic classification of the Mammalia. In (W. P. Luckett & F. S. Szalay, Eds) *Phylogeny of the Primates: A Multidisciplinary Approach,* pp. 21–46. New York: Plenum Press.
Miyamoto, M. M. & Goodman, M. (1986). Biomolecular systematics of eutherian mammals: phylogenetic patterns and classification. *Syst. Zool.* **35,** 230–240.
Novacek, M. J. (1977). Aspects of the problem of variation, origin and evolution of the eutherian auditory bulla. *Mammal Rev.* **7,** 131–149.
Novacek, M. J. (1980). Cranioskeletal features in tupaiids and selected Eutheria as phylogenetic evidence. In (W. P. Luckett, Ed.) *Comparative Biology and Evolutionary Relationships of Tree Shrews,* pp. 35–93. New York: Plenum Press.
Novacek, M. J. (1982). Information for molecular studies from anatomical and fossil evidence on higher eutherian phylogeny. In (M. Goodman, Ed.) *Macromolecular Sequences in Systematic and Evolutionary Biology,* pp. 3–41. New York: Plenum Press.
Novacek, M. J. (1985*a*). Evidence for echolocation in the oldest known bats. *Nature* **315,** 140–141.
Novacek, M. J. (1985*b*). Cranial evidence for rodent affinities. In (W. P. Luckett & J.-L. Hartenberger, Eds) *Evolutionary Relationships Among Rodents: A Multidisciplinary Analysis,* pp. 59–81. New York: Plenum Press.
Novacek, M. J. (1986). The skull of leptictid insectivorans and the higher-level classification of eutherian mammals. *Bull. Am. Mus. nat. Hist.* **183,** 1–112.
Novacek, M. J., McKenna, M. C., Neff, N. A. & Cifelli, R.L. (1983). Evidence from earliest known erinaceomorph basicranium that insectivorans and primates are not closely related. *Nature* **306,** 683–684.
Novacek, M. J. & Wyss, A. (1986). Higher-level relationships of the recent eutherian orders: morphological evidence. *Cladistics* **2,** 257–287.
Parent, J.-P. (1983). Anatomie et valeur systématique de l'oreille moyenne des rongeurs actuels et fossiles. *Mammalia* **47,** 93–122.
Patterson, C. & Rosen, D. E. (1977). Review of ichthyodectiform and other Mesozoic teleost fishes and the theory and practice of classifying fossils. *Bull. Am. Mus. nat. Hist.* **158,** 81–172.
Pettigrew, J. D. (1986). Flying primates? Megabats have the advanced pathway from eye to midbrain. *Science* **231,** 1304–1306.

Pettigrew, J. D. & Cooper, H. M. (1986). Aerial primates: advanced visual pathways in megabats and gliding lemurs. *Soc. Neurosci. Abstr.* **12,** 1035.

Presley, R. (1979). The primitive course of the internal carotid artery in mammals. *Acta anat.* **103,** 238–244.

Rich, T. H. V. & Rich, P. V. (1971). *Brachyerix*, a Miocene hedgehog from western North America, with a description of the tympanic regions of *Paraechinus* and *Podogymnura*. *Am. Mus. Novit.* **2477,** 1–58.

Rose, K. D. (1975). The Carpolestidae, early Tertiary primates from North America. *Bull. Mus. comp. Zool. Harv.* **147,** 1–74.

Rose, K. D. & Gingerich, P. D. (1976). Partial skull of the plesiadapiform primate *Ignacius* from the Early Eocene of Wyoming. *Contr. Mus. Paleont. Univ. Mich.* **24,** 181–189.

Rose, K. D. & Simons, E. L. (1977). Dental function in the Plagiomenidae: origin and relationships of the mammalian Order Dermoptera. *Contr. Mus. Paleont. Univ. Mich.* **24,** 221–236.

Russell, D. E. (1964). Les mammifères paléocènes d'Europe. *Mém. Mus. natn. Hist. nat., Paris, sér. C* **13,** 1–321.

Saban, R. (1963). Contribution à l'étude de l'os temporal des primates. Descriptions chez l'Homme et les prosimiens. Anatomie comparée et phylogénie. *Mém. Mus. natn. Hist. nat., Paris, sér. A* **29,** 1–378.

Sarich, V. M. (1985). Rodent macromolecular systematics. In (W. P. Luckett & J.-L. Hartenberger, Eds) *Evolutionary Relationships Among Rodents: A Multidisciplinary Analysis*, pp. 423–452. New York: Plenum Press.

Savage, D. E. & Russell, D. E. (1983). *Mammalian Paleofaunas of the World.* Reading: Addison-Wesley Publishing Company, Inc.

Schwartz, J. H., Tattersall, I. & Eldredge, N. (1978). Phylogeny and classification of the primates revisited. *Yb. phys. Anthrop.* **21,** 95–133.

Shoshani, J. (1986). Mammalian phylogeny: comparison of morphological and molecular results. *Molec. Biol. Evol.* **3,** 222–242.

Shoshani, J., Goodman, M., Czelusniak, J. & Braunitzer, G. (1985). A phylogeny of Rodentia and other eutherian orders: parsimony analysis utilizing amino acid sequences of alpha and beta hemoglobin chains. In (W. P. Luckett & J.-L. Hartenberger, Eds) *Evolutionary Relationships Among Rodents: A Multidisciplinary Analysis*, pp. 191–210. New York: Plenum Press.

Simons, E. L. (1972). *Primate Evolution: An Introduction to Man's Place in Nature.* New York: Macmillan.

Simons, E. L. & Russell, D. E. (1960). Notes on the cranial anatomy of *Necrolemur. Breviora* **127,** 1–14.

Simpson, G. G. (1935). The Tiffany fauna, Upper Paleocene. 2. Structure and relationships of *Plesiadapis. Am. Mus. Novit.* **816,** 1–30.

Smith, J. D. & Madkour, G. (1980). Penial morphology and the question of chiropteran phylogeny. In (D. E. Wilson & A. L. Gardner, Eds) *Proceedings of the Fifth International Bat Research Conference*, pp. 347–365. Lubbock: Texas Tech Press.

Strickler, T. L. (1978). Functional osteology and myology of the shoulder in the Chiroptera. *Contr. vert. Evol.* **4,** 1–198.

Szalay, F. S. (1969). Mixodectidae, Microsyopidae, and the insectivore-primate transition. *Bull. Am. Mus. nat. Hist.* **140,** 193–330.

Szalay, F. S. (1972). Cranial morphology of the early Tertiary *Phenacolemur* and its bearing on primate phylogeny. *Am. J. phys. Anthrop.* **36,** 59–76.

Szalay, F. S. (1975). Phylogeny of primate higher taxa: The basicranial evidence. In (W. P. Luckett & F. S. Szalay, Eds) *Phylogeny of the Primates: A Multidisciplinary Approach*, pp. 91–125. New York: Plenum Press.

Szalay, F. S. (1977). Phylogenetic relationships and a classification of the eutherian Mammalia. In (M. K. Hecht, P. C. Goody & B. M. Hecht, Eds) *Major Patterns in Vertebrate Evolution*, pp. 315–374. New York: Plenum Press.

Szalay, F. S. & Dagosto, M. (1980). Locomotor adaptations as reflected on the humerus of Paleogene primates. *Folia primat.* **34,** 1–45.

Szalay, F. S. & Decker, R. L. (1974). Origins, evolution, and function of the tarsus in Late Cretaceous Eutheria and Paleocene primates. In (F. A. Jenkins, Jr, Ed) *Primate Locomotion*, pp. 223–259. New York: Academic Press.

Szalay, F. S. & Delson, E. (1979). *Evolutionary History of the Primates.* New York: Academic Press.

Szalay, F. S. & Drawhorn, G. (1980). Evolution and diversification of the Archonta in an arboreal milieu. In (W. P. Luckett, Ed.) *Comparative Biology and Evolutionary Relationships of Tree Shrews*, pp. 133–169. New York: Plenum Press.

Szalay, F. S., Tattersall, I. & Decker, R. L. (1975). Phylogenetic relationships of *Plesiadapis*—postcranial evidence. In (F. S. Szalay, Ed.) *Contributions to Primatology.* Vol. 5, *Approaches to Primate Paleobiology*, pp. 136–166. Basel: Karger.

Van Valen, L. (1965). Treeshrews, primates, and fossils. *Evolution* **19,** 137–151.

Van Valen, L. (1971). Adaptive zones and the orders of mammals. *Evolution* **25,** 420–428.

Van Valen, L. (1979). The evolution of bats. *Evol. Theory* **4,** 103–121.

Wahlert, J. H. (1974). The cranial foramina of protrogomorphous rodents: an anatomical and phylogenetic study. *Bull. Mus. comp. Zool. Harv.* **146,** 363–410.

Wible, J. R. (1984). The ontogeny and phylogeny of the mammalian cranial arterial pattern. Ph.D. dissertation, Duke University.

Wible, J. R. (1986). Transformations in the extracranial course of the internal carotid artery in mammalian phylogeny. *J. vert. Paleont.* **6,** 313–325.
Wible, J. R. (in press). The eutherian stapedial artery: character analysis and implications for superordinal relationships. *Zool. J. Linn. Soc.*
Wible, J. R. & Novacek, M. J. (in press). Cranial evidence for the monophyletic origin of bats. *Am. Mus. Novit.*
Wiley, E. O. (1981). *Phylogenetics: The Theory and Practice of Phylogenetic Systematics.* New York: John Wiley & Sons.
Wilson, J. A. (1966). A new primate from the earliest Oligocene, West Texas, preliminary report. *Folia primat.* **4,** 227–248.
Wood, A. E. (1962). The early Tertiary rodents of the Family Paramyidae. *Trans. Am. phil. Soc.* **52,** 1–261.
Wyss, A. R., Novacek, M. J. & McKenna, M. C. (1987). Amino acid sequence versus morphological data and the interordinal relationships of mammals. *Molec. Biol. Evol.* **4,** 99–116.
Zeller, U. (1986). Ontogeny and cranial morphology of the tympanic region of the Tupaiidae, with special reference to *Ptilocercus. Folia primat.* **47,** 61–80.

Jeffrey H. Schwartz

Department of Anthropology, University of Pittsburgh, Pittsburgh, PA 15260, U.S.A.

Ian Tattersall

Department of Anthropology, American Museum of Natural History, New York, NY 10024, U.S.A.

Received 30 November 1986
Revision received 24 June 1987
and accepted 25 June 1987
Publication date November 1987

Keywords: adapids, phylogeny, primates, prosimians, strepsirhines, systematics.

Tarsiers, adapids and the integrity of Strepsirhini

Although strepsirhine primates can be described by their narial configuration, this and most other definable features are probably primitive retentions; only the development of a grooming claw of the second pedal digit and of a toothcomb (the latter of which has been lost in *Daubentonia*) emerge as potential apomorphies of the group. Within this assemblage lemurids, *Lepilemur*, the indriids, and *Daubentonia* can be argued to constitute a monophyletic group whose relationships cladistically are in the sequence listed; Lemuridae and Indriidae can themselves be delineated as monophyletic groups. The remaining strepsirhine primates—the cheirogaleids, galagids, and lorisids—also appear to constitute a definable clade, with the former group representing the sister taxon of the latter two families; each family can be united on the basis of distinct synapomorphies. Although there are features—especially of the ear region—which present themselves as potentially reflective of the sister relationship of *Tarsius* + Anthropoidea, other characters, including the possession of the grooming claw, are suggestive of an alternative scheme: *Tarsius* may be the sister of the extant lorisiform group, thereby reconstituting, albeit in a novel form, the primate suborder Prosimii. It also appears that fossil "tarsioids" may in fact be more closely related to the extant lorisiforms than to *Tarsius*. A reconsideration of the so-called fossil lemurs, the adapids, leads to the conclusion that *Adapis*-like primates are a clade apart from *Pelycodus*, *Notharctus*, *Smilodectes* and their most immediate relatives, and may themselves constitute a clade that is related as the primitive sister to all other "prosimians" by virtue of the development of the so-called free intrabullar tympanic ring.

Journal of Human Evolution (1987) **16,** 23–40

Introduction

The decade of the 1970s witnessed a shift in consensus away from the subdivision of Primates into Prosimii and Anthropoidea and towards the division of the order into Strepsirhini and Haplorhini (e.g., Luckett, 1975; Szalay, 1975). Crucial to this shift, of course, was the reinterpretation of the relationships of the enigmatic tarsier. During the same period, doubt began to be cast on the neat geographical division of Strepsirhini into Malagasy and non-Malagasy forms (e.g., Tattersall & Schwartz, 1974). Most recently, the concept that the Eocene adapids were in some way broadly ancestral as a group to the extant strepsirhine primates has been subject to question (e.g., Schwartz & Tattersall, 1985). It may thus be appropriate at this juncture to review the evidence for these points of view, and to survey the morphological evidence for the relationships among and within these groups of primates. With regard to Haplorhini, we have referred principally to such recent reviews as Aiello (1986) and MacPhee & Cartmill (1986), which provide summaries and tabulations of features in apparent support of this clade.

Relationships among the strepsirhines

The first question to be addressed in this context is whether Strepsirhini (i.e., "lemurs" + "lorises") does indeed constitute a coherent clade, and if so, what are the relationships within the clade? Based on the analysis of several studies investigating various different aspects of morphology (Bugge, 1974; Jouffroy *et al.*, 1984; Luckett, 1974; Cartmill, 1975;

0047 2484/87/010023 + 18 $03.00/0

Ford, 1980; Hofer, 1980; MacPhee & Cartmill, 1986; Stephan, 1984; Schwartz & Tattersall, 1985), we have compiled a table of some 45 characters and states (Table 1) that have served in previous studies as the basis for establishing hypotheses of relationships within the group, and which are employed here once again for that purpose. Our hypothesis of relationships is represented in Figure 1.

The most obvious synapomorphy uniting the lemurs and lorises as a whole is the presence of a grooming claw on the second digit of the foot. Of the taxa united by this feature, all but one extant form, *Daubentonia*, possess a toothcomb. Unquestionably, the loss of the toothcomb and the acquisition of continuously growing, laterally compressed anterior teeth represent an autapomorphy of this genus. The lemurs and lorises also share the "strepsirhine" configuration of the nares, i.e. the nostrils slit or creased laterally, with the presence of a rhinarium, a broad internarium, and unfused median nasal prominences. However, a broader comparison among mammals reveals these characters to be primitive retentions, and thus of no utility in recognizing this clade. Similar primitive retentions for the group include the presence of a prepollex, an elongated fourth pedal digit, moderately developed occipital extension of the cerebral hemispheres, short and broad olfactory

Table 1 **Distribution of various morphologies among extant strepsirhines and *Tarsius***

	Lemur's	*Lepi.*	*Daub.*	Indriids	Cheiro's	Galagids	Lorisids	*Tarsius*
Grooming claw:								
2nd pedal digit	×	×	×	×	×	×	×	×
3rd pedal digit	0	0	0	0	0	0	0	×
Toothcomb	×	×	0	×	×	×	×	?
Ectotympanic:								
aphaneric	×	×	×	×	×	—	—	×
phaneric	—	—	—	—	—	×	×	—
fused/unfused	unfus.	unfus.	unfus.	unfus.	unfus.	fus.	fus.	unfus.
tubular	—	—	—	—	—	—	×	×
Position carotid f.	post-lat.	post-lat.	post-lat.	post-lat.	post.	post-med.	post-med.	cent.
Promontory artery	sm.	sm.	sm.	sm.	sm.	sm.	lg.	lg.
Stapedial artery	lg.	vest.	lg.	lg.	vest.-0	vest.-0	vest.-0	vest.-0
Ascending pharyngeal artery (pharyngeo-carotid anastomosis)	0	0	0	0	×	×	×	0
Foramen lacerum exposed	0	0	0	0	×	×	×	0
Middle meningeal a. fed via ophthalmic a.	0	0	0	0	×	×	×	0
Anterior accessory cavity	0	0	0	0	0	0	×	×
Tympanic cavity	lg.	lg.	lg.	lg.	lg.	sm.	sm.	sm.
Annular bridge	×	×	×	×	×	0	0	0
Mastoid cavity	sm.	sm.	sm.	sm.	lg.	lg.	lg.	sm.
Epitympanic recess: complex	0	0	0	0	×	×	×	0

Table 1 *continued*

	Lemur's	*Lepi.*	*Daub.*	Indriids	Cheiro's	Galagids	Lorisids	*Tarsius*
Pterygoid plates medial plates	sub= sep.	sub= sep.	sub= sep.	sub= sep.	sub= sep.	sub= sep.	sub= sep.	lat.>>med. fus.
Glenoid fossa	shal./ flat	shal./ flat	shal./ flat	shal./ flat	shal./ flat	shal./ flat	shal./ flat	deep pock.
Orbital frontation	br.	br.	br.	br.	mod.	mod.	mod.-mk.	mk.
Rostral elongation	0	0	0	0	mod.	mod.	mod.-mk.	mod.
Premaxillary margin	undist.	undist.	undist.	undist.	dist.	dist.	dist.	dist.
Ethmoid exposure	0 (*Hapa.*)	×	0	occ.	×	×	×	×
Palatine-frontal contact	× (*Hapa.*)	0	0	0	0	0	0	×
Maxilloethmoid contact	0	0	0	0	× (*Cheiro?*)	×	×	×
Nares	strep.	strep.	strep.	strep.	strep.	strep.	strep.	strep.
Internarium	br.	br.	br.	br.	br.	br.	br.	ex.br.
Median nasal prominences	unfus.	unfus.	unfus.	unfus.	unfus.	unfus.	unfus.	fus.
Olfactory peduncle	sh.br.	sh.br.	sh.br.	sh.br.	sh.br.	sh.br.	sh.br.	sh.br.
Retinal fovea	poor	poor	poor	poor	poor	poor	poor	poor
Calcaneus	sh.	sh.	sh.	sh.	lng	lng (*G.crass.*)	sh.	ex.lng
Navicular	sh.	sh.	sh.	sh.	lng	lng (*G.crass.*)	sh.	ex.lng
Trochlear facet	sh.br.	sh.br.	sh.br.	sh.br.	nar.lng	nar.lng	nar.lng	sh.br.
Tibia-fibula	unfus.	unfus.	unfus.	unfus.	fus.	fus.	unfus.	fus.
3rd finger	mod.	mod.	mod.	mod.	mod.	mod.	mod.	lng
4th toe	lng	lng	lng	lng	lng	lng	lng	lng
Articular surface mandib. condyle	str.	"lip"	arc.	arc.	str.	str.	str.	comp.
Paraconid	0 (*Hapa.*)	0	0	0	0	0	0	M_{1-3}
Paracristid	arc.	ant.	"ant."	ant.	low	low	low	ant.
M_1 trigonid	lng	open	"open"	open	comp.	comp.	comp.	open
Protoconid-metaconid: connection height	crest low	sep. low	"sep." low	sep. low	melded tall	melded tall	melded tall	melded tall
Hypoconid	rnd	rnd	rnd	rnd	ang.	ang.	ang.	ang.
Prehypocone crista, M^{1-2}	0	0	0	0	×	×	×	×

Key to abbreviations: Lemur's = Lemurids; *Lepi.* = *Lepilemur*; *Daub.* = *Daubentonia*; Cheiro's = Cheirogaleids; f. = foramen; × = present; 0 = not present; poss. = possible; — = not applicable; fus. = fused; unfus. = unfused; shal. = shallow; pock. = pocket; lat. = lateral; med. = medial; cent. = central; post = posterior; sm. = small; lg. = large; vest. = vestigial; br. = broad; mod. = moderate; mk. = marked; dist. = distended; undist. = undistended; occ. = occasional; strep. = strepsirhines; sh. = short; lng = long; nar. = narrow; str. = straight; arc. = arcuate; comp. = compressed; ant. = anterior; sep. = separate; rnd = round; ang. = angular; ex. = extra; taxa in brackets = exceptions; *Hapa.* = *Hapalemur*; *G.crass.* = *Galago crassicaudatus*; *Cheiro* = *Cheirogaleus*.

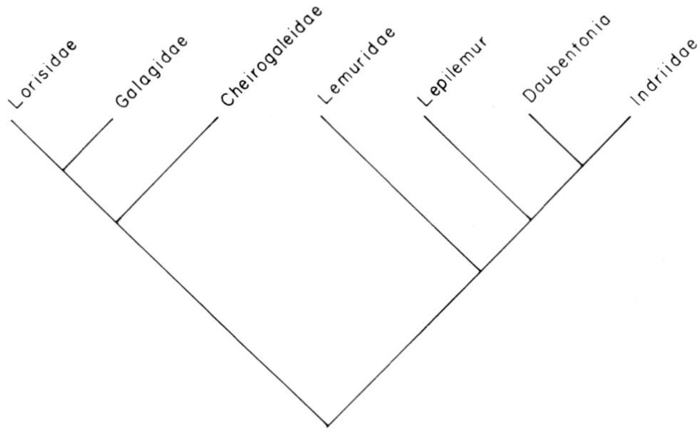

Figure 1. Hypothesis of relationships among the major groups of extant strepsirhine primates.

peduncle, poorly developed retinal fovea, a capacious nasal fossa, and, in placentation, the disruption of Rauber's layer during amniogenesis (which is completed by folding), the presence during the development of an epitheliochorial placenta of a choriovitelline stage, and the presence of a large free yolk sac.

Within this wider clade, a subgroup is characterized by the development of an anterior carotid or ascending pharyngeal (see below) artery; anterior palatine fenestrae that separate the upper incisors; the development of prehypocone cristae, at least on M^1; lower molar protoconids and metaconids that are melded at their bases; a paracristid that rings the base of these cusps, truncating the trigonid basin; the possession of a narrow and elongate trochlear facet of the calcaneus; reduction/loss of the stapedial artery in the adult; exposure of the foramen lacerum, which admits a pharyngeocarotid anastomosis; and a middle meningeal artery that is supplied primarily by the ophthalmic artery. These features unite the lorisiforms, i.e. the lorisids, galagids, and cheirogaleids. Cartmill (1975) and MacPhee & Cartmill (1986) suggested that the vessel in genus *Lemur* identified by the Saban (1963) as the "rameau tubaire", may be the homologue of the "ascending pharyngeal" (better known, and more appropriately identified, by its earlier appellation of "anterior carotid") of lorisiforms; this small vessel, it should be noted, does not proceed to form a pharyngeocarotid anastomosis, as it does in the latter. Such an anastomosing vessel has not been identified in any non-cheirogaleid lemuriform and we doubt that this claimed homology can be regarded as securely established. In turn, the lorisids and galagids are united by a suite of synapomorphies that includes a tympanic ring fused to the bullar edge, and the development on the lower molars of tall protoconids and metaconids and a deep hypoflexid notch that tends to isolate the protoconid.

A clade Lemuriformes, containing Lemuridae, Megaladapidae (as represented among the living forms by *Lepilemur*), Daubentoniidae and Lemuridae, is united by a set of dental synapomorphies that include compressed upper molar cusps bearing distinct crests, a broadening and lowering of the upper molar protocone in the sequence M^{1-3}, some distension posteriorly of the metaconid in the lower molars, and anterior premolars with anterior elongation. Generally, the presence of a large stapedial branch and a small promotory artery has been regarded as a primitive character retained by these primates; it is possible, however, especially in view of the presence in *Plesiadapis* of the reverse condition, that it is the latter which is derived. In that case, the lemuriform condition in this character would represent a synapomorphy for the group. However, sufficient uncertainty exists about the polarity of this morphocline that at present we prefer to leave this entire character complex aside when constructing hypotheses of relationship.

Within Lemuriformes, *Lepilemur* (Megaladapidae) is united with *Daubentonia* plus the indriids by the development of a secondary articulation on the posterior aspect of the mandibular condyle, de-emphasis of the postprotocristae on M^{1-3}, broad trigons on M^{1-3}, distended para- and metastylar regions on M^{1-2}, elongate M_1 with marked posterior distension of the metaconid, and a lower posterior premolar that is deeply excavated distally. *Daubentonia* is, of course, highly autapomorphic, but this form shares with the indriids a rounded, globular cranium with deepened splanchnocranium, a deep mandible with expanded gonial region and condyles rounded in the coronal plane, M^{1-2} quadrate with broadened hypocone, and enlarged para- and metaconule and mesostylar regions.

The relationships of *Tarsius*

Two primary points of view are current on the affinities of the tarsier. Either this enigmatic form is the sister of the lemurs + lorises in the suborder Prosimii, or it is the sister of the anthropoids in the suborder Haplorhini (see review by Aiello, 1986). It may be worthwhile, however, to consider alternative possibilities, which are essentially two: either *Tarsius* is the sister of Strepsirhini + Anthropoidea, or it nests within one of these groups. To sustain the first of these possibilities requires a complex set of assumptions uniting *Tarsius* with a host of supposed fossil relatives (see below), and the ensemble with the plesiadapiforms on the basis of the shared possession of enlarged anterior teeth (see review by Schwartz, 1984). A more robust theory of relatedness can, however, be constructed by considering the other alternative (Figure 2).

While Prosimii as conventionally constituted was essentially based on notions of grade, more recent arguments for the existence of a haplorhine clade have been founded on more substantial considerations. Initial features brought to bear focussed on aspects of the narial region—primarily the aboral rounding of the nares (Pocock, 1918)—to which were later

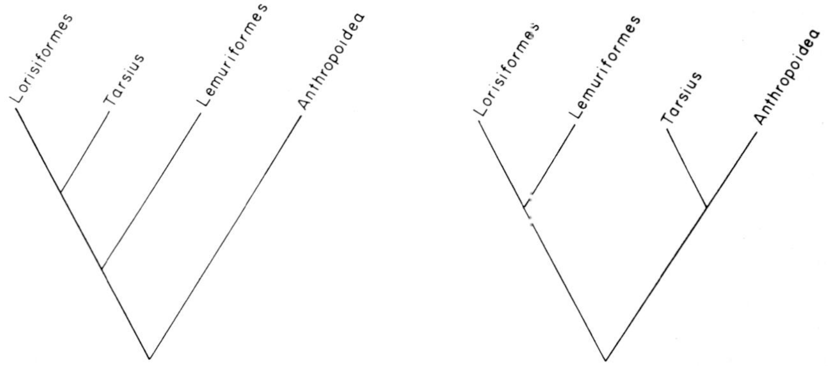

Figure 2. Two possible hypotheses of the relationships between *Tarsius* and other primate groups. Our preferred scheme is shown left.

added features such as fusion of the upper lip and the absence of a moist, naked rhinarium (Hill, 1955). Most recently, the development of hemochorial placentation and attendant aspects of amniogenesis and placentation (Luckett, 1974), as well as features of the orbit, nasal capsule and auditory region (e.g., Aiello, 1986; MacPhee & Cartmill, 1986), have been cited as further support for the reality of Haplorhini. With regard to the configuration of the nostrils, Hofer (1980) has recently pointed out that in *Tarsius* as well as in some platyrrhines one finds variably developed creases reminiscent of a strepsirhine condition. Further, while Hofer (1980) confirms Hill's (1955) conclusion that *Tarsius* resembles anthropoids in possessing a fused upper lip, he points out that there are two ways in which this condition can occur, namely the coalescence very early in development of the maxillary and median nasal processes, and the developmentally later gradual obliteration of the median rhinarial fissure. The possibility that syncheilism has been achieved by different means in different primate groups displaying this condition is suggested by the findings of Hershkovitz (1977) and Hofer (1977) of vestigial median furrows, with sinus hairs, in various platyrrhines. Szalay (1977), moreover, has pointed out that other mammals such as equids and bovids have fused median nasal prominences and also lack a moist, naked rhinarium. As in *Tarsius* and anthropoids, the fused upper lips of these mammals bear hairs. In some cases (e.g. various platyrrhines), these are sinus hairs; in others, e.g. apes and humans, they are not. *Tarsius*, unfortunately, is insufficiently known in this particular respect, but overall Hofer (1980, p. 247) was confident enough to state that on the evidence of the nasal region tarsiers "occupy a place remote from Anthropoidea".

As regards the nasal capsule, reduction in size has been cited as a synapomorphy of *Tarsius* and anthropoids (Cave, 1973), but as Starck (1975) has noted, reduction of the nasal capsule in anthropoids is absolute, whereas in *Tarsius* this region is diminished in size by the encroachment of the greatly expanded orbital region, which factor also accounts for the absence of the olfactory recess. Similarly, the presence of an "apical" interorbital septum has been cited (Cartmill, 1972; Cave, 1973; Aiello, 1986) as indicative of the tarsier's affinity with anthropoids. Clearly, however, the extreme extent of the septum in *Tarsius* is due to the great expansion of the orbits (Starck, 1975); indeed, Cave (1973) points to deficiency in the septum as a consequence of orbital size.

In a series of valuable contributions, Cartmill (1975, 1978, 1981) has provided much detail on the development and morphology of the medial orbital wall in various primates. He has argued that this area in *Tarsius* shows a conformation intermediate between the conditions shown by strepsirhines and anthropoids. Principally, he has pointed to the expansion of the alisphenoid contribution in both *Tarsius* and the latter. However, the hypertrophied orbit of *Tarsius*—which, as expanded as it is, merely cups shallowly the enormous eyeball (Castenholz, 1984)—is in fact a highly autapomorphic structure in most respects, including, most strikingly, the extreme expansion of the maxillary, frontal and (superiorly) the zygomatic components (Cartmill, 1981, p. 246). Even if one accepts Cartmill's arbitary reduction (not scaled for size) of the area of the *Tarsius* medial orbital wall appropriate for comparison with other primates, the relative proportions of the contributory elements still point essentially towards the autapomorphic nature of this structure as a whole in the tarsier. Thus, while expansion of the alisphenoid must remain a potential synapomorphy uniting the tarsier with anthropoids (although some of the latter, at least, would appear to lack this expansion: see illustrations of *Alouatta* and *Cebus* in Cartmill, 1978, p. 102), we do not feel that this feature can be employed with much confidence in phylogenetic reconstruction. Essentially, we must concur with Starck's (1984) implicit conclusion that the development of the interorbital area of *Tarsius* is effectively unique.

An aspect of the anatomy of the eye that has recently been used to associate the tarsier with the anthropoids (Aiello, 1986) is retinal morphology, including the supposed presence of a macula lutea as reported by Woollard (1926). On the basis of further materials, however, Kollmer (1930) concluded that the "macula" found by Woollard was an artifact of fixation, and that the retina of the tarsier essentially resembled that of *Loris* (Wolin & Massopust, 1970). On the basis of more recent reports (e.g., Castenholz, 1965, 1984; Rohen, 1966; Wolin & Massopust, 1970) it is arguable whether or not a fovea exists (or if it does, if it occurs consistently) in the pure rod retina of the tarsier. Following a review of the available evidence, including his own, Castenholz (1984) has concluded that attributes of visual morphology and physiology of the tarsier are of no value as taxonomic indicators.

MacPhee & Cartmill (1986) have recently provided a detailed consideration of the structures of the primate cranial base. Of seven potential synapomorphies between the tarsier and the anthropoids, they felt most confident of four. These are: absence of an annular bridge, possession of a small tympanic cavity, the perbullar (as opposed to transpromontorial) path of the internal carotid artery, and the development of an anterior accessory cavity of the middle ear, as a diverticulum of the auditory tube. One might also include in this category the supply of the middle meningeal artery primarily by the maxillary artery. Among these, however, the absence of an annular bridge and a small tympanic cavity are also characteristic of lorisids and galagids, leaving only three major features of this region uniquely shared by tarsiers and anthropoids. MacPhee & Cartmill suggest that the vestigial stapedial and unreduced promontory arteries may constitute an additional synapomorphy of the latter. However, this appears in fact to be a combination of two separate characters, since severe reduction of the stapedial artery is also characteristic of cheirogaleids, galagids and lorisids, in which the promontory artery is small or lacking altogether. MacPhee & Cartmill's data show, moreover, three potential synapomorphies shared uniquely between lorisiforms and anthropoids: development of a complex epitympanic recess, a large mastoid cavity, and a posteromedial point of penetration of the auditory region by the carotid artery. In sum, the phylogenetic signal of characters of the auditory region is far from clear.

Table 2 Distribution of various morphologies among extant strepsirhines, *Tarsius* and anthropoids

	Lemuriforms	Lorisiforms	*Tarsius*	Callitrichids	Cebids	Catarrhines
Grooming claw:						
2nd pedal digit	×	×	×	0	0	0
3rd pedal digit	0	0	×	0	0	0
Toothcomb	× (*Daub.*)	×	?	0	0	0
Ectotympanic:						
apheneric	×	cheiro's	×	—	—	—
phaneric	—	×	—	×	×	×
fused/unfused	unfus.	fus. (cheiro's)	unfus.	fus.	fus.	fus.
tubular	—	—	×	—	—	×

Table 2 *continued*

	Lemuriforms	Lorisiforms	*Tarsius*	Callitrichids	Cebids	Catarrhines
Position carotid f.	post-lat.	post-med. (cheiro's)	cent.	post-med.	post-med.	post-med.
Path of internal carotid artery	transprom.	transprom.	perbull.	perbull.	perbull.	perbull.
Promontory artery	sm.	sm.-0	lg.	lg.	lg.	lg.
Stapedial artery	lg.	vest.-0	vest.-0	vest.-0	vest.-0	vest.-0
Ascending pharyngeal artery (pharyngeo-carotid anastomosis)	0	×	0	0	0	*Homo* (?)
Middle meningeal a.: primary supply	stap.	ophthal.	max.	max.	max.	max.
Anterior accessory cavity	0	×	×	×	×	×
Foramen lacerum exposed	0	×	0	0	0	0 (*Homo, Pongo*)
Tympanic cavity	lg.	sm. (cheiro's)	sm.	sm.	sm.	sm.
Annular bridge	×	0 (cheiro's)	0	0	0	0
Mastoid cavity	sm.	lg. (cheiro's)	lg.	lg.	lg.	lg.
Epitympanic recess: complex	0	× (cheiro's)	0	×	×	×
Orbital frontation	br.	mod.-mk.	mk.	mod.-mk.	mod.	mod.
Rostral elongation	0	mod.-mk.	mod.	0	0	0
Nasal fossa	unred.	unred.	comp. red.	abs. red.	abs. red.	abs. red.
Pterygoid plates medial plates	sub= sep.	sub= sep.	lat.>>med. fus.	sub= sep.	sub= sep.	sub= sep.
Glenoid fossa	shal./ flat	shal./ flat	deep pocket	shal./ flat	shal./ flat	shal./ flat
Premaxillary margin	undist.	dist.	dist.	undist.	undist.	undist.
Ethmoid exposure	0 (indriids, *Hapa.*, *Lepi.*)	×	×	×	×	×
Palatine-frontal contact	0 (*Lemur*)	0	×	×	×	×
Nares	strep.	strep.	strep.	strep.	hap.	hap.
Internarium	br.	br.	ex.br.	br.	br.	nar.
Internarial hair	0	0	×	sinus	sinus	×(?)
Median nasal prominences	unfus./ mdn.fiss.	unfus./ mdn.fiss.-sulc	fus.	fus./ mdn.gr.	fus./ mdn.gr.	fus.

Table 2 *continued*

	Lemuriforms	Lorisiforms	*Tarsius*	Callitrichids	Cebids	Catarrhines
Vibrissae	×	×	×	0	0	0
Rauber's layer	lost	lost	lost	intact	intact	intact
Amniogenesis	fold.	fold.	cav.+fold.	cav.	cav.	cav.
Blastocyst implant. trophoblast orientation	paraembryo. orthomes.	paraembryo. orthomes.	paraembryo. mesomet.	embryo. orthomes.	embryo. orthomes.	embryo. orthomes.
Placentation	diff.epi.chorio. lg.ylk.sc.sm. all.	diff.epi.chorio. lg.ylk.sc.sm. all. (*M.mur.*,*G.demi.*)	hem.sm.ylk. sc.mes.bdy	hem.sm.ylk. sc.mes.bdy	hem.sm.ylk. sc.mes.bdy	hem.sm.ylk. sc.mes.bdy
Olfactory peduncle	sh.br.	sh.br.	sh.br.	lng.nar.	lng.nar.	lng.nar.
Retinal fovea	poor	poor	poor	well-dev.	well-dev.	well-dev.
Calcaneus	sh.	lng (lorisids, *G.crass.*)	ex.lng	sh.	sh.	sh.
Navicular	sh.	lng (lorisids, *G.crass.*)	ex.lng	sh.	sh.	sh.
Trochlear facet	sh.br.	nar.lng	sh.br.	sh.br.	sh.br.	sh.br.
Tibia-fibula	unfus.	fus. (lorisids)	fus.	unfus.	unfus.	unfus.
3rd finger	mod.	mod.	lng	lng	lng	lng
4th toe	lng	lng	lng	mod.	mod.	mod.
Articular surface, mandib. condyle	str., "lip" or arc.	str.	comp.	str.	str.	str.
Paraconid	0 (*Hapa.*)	0	M_{1-3}	0	0	0
Paracristid	arc.or ant.	low	ant.	arc.or low	arc.,low or 0	0
Protoconid-metaconid connection	crest (*Lepi.*,*Daub.*,*indriids*)	melded	melded	sep.	sep.	sep. (cerco's)
height	low	tall	tall	low	low	low
Hypoconid	rnd	ang.	ang.	rnd	rnd	rnd
Prehypocone crista, M^{1-2}	0	×	×	0	0	0

Key to abbreviations: cheiro's = Cheirogaleids; f. = foramen; × = present; 0 = not present; — = not applicable; fus. = fused; unfus. = unfused; post = posterior; ant. = anterior; lat. = lateral. med. = medial; cent. = central; transprom. = transpromontorial; perbull; = perbullar; sm; = small; lg. = large; vest. = vestigial; stap. = stapedial; ophthal. = ophthalmic; max. = maxillary; br. = broad; mod. = moderate; mk. = marked; unred. = unreduced; comp.red. = reduced via compression; abs.red. = absolutely reduced; shal. = shallow; dist. = distended; undist. = undistended; occ. = occasionally; strep. = strepsirhine; hap. = haplorhine; mdn. = median; fiss. = fissure; sulc. = sulcus; gr. = groove; implant. = implantation; paraembryo. = paraembryonic; embryo. = embryonic; orthomes. = orthomesometrial; mesomet. = mesometrial; diff.epi. = diffuse epitheliochorial; chorio. = choriovitelline placenta; ylk.sc. = yolk sac; all. = allantoic vesicle; hem. = discoidal hemochorial; mes.bdy = mesodermal body stalk; sh. = short; lng = long; nar. = narrow; well-dev. = well-developed; str. = straight; arc. = = arcuate; comp. = compressed; ant. = anterior; sep. = separate; rnd = round; ang. = angular; ex. = extra; taxa in brackets = exceptions; *Hapa.* = *Hapalemur*; *Lepi.* = *Lepilemur*; *Daub.* = *Daubentonia*; *M.mur.* = *Microcebus murinus*; *G.demi.* = *Galago demidovii*; *G.crass.* = *Galago crassicaudatus*; cerco's = cercopithecids.

Luckett (e.g., 1974, 1976) has argued repeatedly that features of amnion formation and placentation convincingly unite *Tarsius* and Anthropoidea. Among the features he cites are the development of a mesodermal body stalk, a rudimentary allantoic vesicle, a small free yolk sac, the presence of a primordial amniotic cavity, lack of a choriovitelline placental stage, and the development of a discoidal hemochorial placenta. Differences of *Tarsius* from anthropoids, as in the implantation of the blastocyst to the mesometrial pole of the uterine endometrium, and the shift during amniogenesis from cavitation to folding, have been attributed to the tarsier's possession of a bicornuate, rather than simplex, uterus. However, there are other mammals (e.g., dasypodids, edentates, and various rodents) with hemochorial placentae that have bicornuate uteri but which nevertheless develop "anthropoid" characteristics of blastocyst implantation, amniogenesis and placentation (Luckett, 1974). Moreover Luckett (1975, p. 179) has pointed out that, in the development of a hemochorial placenta, the precocious establishment of a chorioallantoic placenta bypassing the choriovitelline stage, the rudimentary development of an allantoic diverticulum, and the precocious differentiation of a mesodermal body stalk, are all perfectly correlated. They thus form a single functional complex rather than an aggregate of independent characters. Further, since as Luckett (1975, p. 179) argues, "amniogenesis by cavitation [as seen in anthropoid primates] *necessitates* [emphasis added] the development of a primordial amniotic cavity", the development of this cavity "in concert with definitive amniogenesis by folding, as in *Tarsius*", cannot be regarded as a homologous characteristic, and certainly not as a synapomorphy of haplorhines. In sum, it appears that *Tarsius* begins its course of placental and amnion formation in a manner that is totally dissimilar from that of anthropoids, and that, like a variety of other mammals, it converges on the anthropoid condition purely out of functional necessity. Moreover, *Tarsius* is autapomorphic among primates in beginning amniogenesis by cavitation but completing this process by folding. In support of the tarsier/anthropoid hypothesis, Aiello (1986) cited Leutenegger's (1973) finding that (in a sample of one) *Tarsius syrichta* showed a relatively high birth weight. Leutenegger himself, however, considered this simply to be a correlate of hemochorial placentation, already discussed.

A feature that unites cheirogaleids, galagids, lorisids and *Tarsius* is the presence on M^{1-2} of a prehypocone crista (Schwartz, 1984). All, moreover, share a M_1 protoconid-metaconid melded at the base and forming a steep wall facing on the talonid, an angular hypoconid on M_{1-2} that is distended buccally, a premaxillary alveolar margin that is elongated downward, and a marked degree of orbital frontation and relative enlargement. Further, scanning electron micrographs of the anterior lower teeth of *Tarsius pumilus* (Musser & Dagosto, 1987) reveal that "I_1" morphology in this species matches that of the lateral tooth of the strepsirhine toothcomb. In another anatomical region, ankle morphology—specifically, the calcaneal and navicular elongation that exists in cheirogaleids and most galagids—hints at an affinity of *Tarsius* with the lorisoid group. Similarly, the highly distinctive grooming claw, whose possession unites all living strepsirhines—and perhaps also adapids too (von Koeningswald, 1979), is also found in *Tarsius*. Tarsiers actually possess two grooming claws, on the second and third pedal digits; the latter feature is plausibly interpreted as an autapomorphy added to the synapomorphy with strepsirhines. In addition, distal tibio-fibular fusion exists in *Tarsius* as well as in the comparably-sized lorisiform primates. Molecular data may point in a similar direction: although Aiello (1986) cites the immunodiffusion studies of Dene *et al.* (1976) as demonstrating the unity of

Tarsius and Anthropoidea, the same group (Baba *et al.*, 1982) later concluded from their studies of hemoglobin sequences that their data yield three phylogenies of equal likelihood, of which *Tarsius* + Anthropoidea is one, but *Tarsius* + Strepsirhini is another. Finally, a recent preliminary chromosomal study by Poorman *et al.* (1985) demonstrates an astonishing amount of chromosomal homology between *Tarsius* and, especially *Galago crassicaudatus*. Hopefully, broader comparative analyses will elucidate the phylogenetic significance of this striking comparison.

To summarize, the only characters uniquely shared by *Tarsius* and Anthropoidea are the perbullar course of the carotid artery, development of an anterior accessory cavity of the auditory bulla, lack of a rhinarium, fusion of the upper lip, a hemochorial placenta, and (possibly) the presumed presence of a long alisphenoid component in the medial orbital wall. As discussed at various points above, however, we find only the first two of these features potentially convincing. On the other hand, *Tarsius* shares the grooming claw and certain other characters with all strepsirhines, and shares various features of the dentition, premaxilla and orbit specifically with lorisiforms. Unsurprisingly when we are dealing with so highly autapomorphic a form as the tarsier, there is no unequivocal case to be made either for a sister relationship with the anthropoids or the lorisiforms—although the case for a clade of *Tarsius* + Strepsirhini as a whole, an earlier orthodoxy, is weak indeed. However, one might be justified in leaning towards the relatedness of *Tarsius* with the lorisiforms because this hypothesis involves fewer features known to have evolved separately in other primate or mammal groups, and because it is supported by the various suggestive, if individually inconclusive, features mentioned above. If the relationships of *Tarsius* do indeed lie with the lorisiforms, then of course its ancestor must have possessed a toothcomb—a possibility reinforced, as noted above, by the morphology of the tooth generally identified as the lower incisor in *Tarsius pumilus* (see Musser & Dagosto, 1987, fig. 21). This tooth has all the morphological features, if not the complete procumbent orientation, of the lateral tooth of the strepsirhine toothcomb. As discussed in the next section, loss or modification of this latter structure is not without parallel amongst strepsirhines.

Adapids and strepsirhines

As noted earlier, a tacit consensus exists that adapids, in a general sense, constitute a monophyletic group that is broadly ancestral to the strepsirhines of today. In order to demonstrate this, however, it is necessary to adduce features that are shared uniquely by forms as diverse as *Adapis* and *Pelycodus*, as well as to find at least one synapomorphy shared by the fossil forms with extant taxa. The latter would seem to be demonstrated rather simply by the similarities of the configuration and relations of the inflated petrosal bulla, and the tympanic ring it surrounds, that are seen in both the Eocene taxa and various modern "lemurs". However, while this "lemur-like" bulla may well prove to be a significant taxonomic indicator at the level of the last common ancestor of all such taxa, it cannot also verify the monophyly of Adapidae.

Elsewhere (Schwartz & Tattersall, 1985, and references therein) we have considered in some detail the European forms which over the years have been assigned to Adapidae, and have concluded that few share any of the distinctive features of *Adapis* itself, hence any characters that could be regarded as apomorphic for a family Adapidae. We thus reduced this taxon to the well documented fossil genera *Adapis, Leptadapis*, and *Paradapis*, and to a couple of less well known forms. Based on the more complete specimens of *Adapis* and *Leptadapis*, features characteristic of Adapidae include submolariform upper and lower last premolars, quadrate M^{1-2}, with M^{1-3} bearing well developed hypocones; ringing of upper and lower cheek teeth by cingula; a non-caniniform ("premolariform") and occlusally rotated lower canine; distension of the metaconid into a metastylid on the last lower premolar and M_{1-3}.

Other "adapids" of the holarctic Eocene include *Notharctus* and its allies (e.g., *Cercamonius*), *Pelycodus* and its allies (e.g., *Cantius, Protoadapis*), and *Smilodectes* and its allies (e.g., *Washakius*—see Schwartz, 1984, 1986). *Smilodectes* and *Notharctus* are generally believed to be descended independently from *Pelycodus* (e.g., Gingerich, 1979), and the group is certainly linked by the common possession of protocone folds on M^{1-3}. Of the three, *Smilodectes* would otherwise appear to have the best claim to be related to *Adapis* and its allies; shared apparent synapomorphies include the quadrate form of the upper molars, the compression of the upper molar buccal cusps, and the accentuation of lower molar crests including the somewhat anteriorly-directed paracristid on M_1. However, various other dental features of *Smilodectes*, including the emphasis on cusps as opposed to crests in the cheek teeth, and the presence of what appears to be a paraconid on M_1, argue against a close relationship with the adapids proper. In any event, more cogent arguments can be made for pairing *Smilodectes* with *Pelycodus* (both have arcuate M_1 cristids obliquae which meet the metaconid) or *Notharctus* (stout cresting on the lower last premolar; development of entoconid notch on M_{1-2}, development of upper molar mesostyles, and expansion of hypocone region on M^{1-2}). We therefore conclude at present that *Smilodectes, Notharctus* and *Pelycodus* together form a more plausible clade than any one of them does with *Adapis* or its allies.

If, then, there exist an *Adapis* clade and a *Pelycodus* clade, what are the affinities of each within the wider setting of Primates? In the course of describing features of European Eocene forms that seemed to distinguish *Adapis* and its close relatives from the *Pelycodus* group (Schwartz & Tattersall, 1985), we delineated a set of characters that included spatulate and mesially expanded upper incisors, cheek teeth with compressed buccal cusps and emphasized stylar regions, quadrate and quadricuspid M^{1-2} with rather broadly

Table 3 Distribution of various morphologies among extant strepsirhines and "adapids"

	Lemur's	*Lepi.*	Indriids	Loris's	*Adapis*	*Pelyc.*	*Noth.*	*Smilo.*
Grooming claw: 2nd pedal digit	×	×	×	×	poss.	unk.	unk.	unk.
Toothcomb	×	×	×	×	?	0	0	0
Ectotympanic: aphaneric	×	×	×	—	×	×	×	×
phaneric	—	—	—	× (cheiro's)	—	—	—	—
fused/unfused	unfus.	unfus.	unfus.	fus. (cheiro's)	unfus.	unfus.	unfus.	unfus.
Position carotid f.	post-lat.	post-lat.	post-lat.	post-med. (cheiro's)	post-lat.	post-lat.	post-lat.	post-lat.

Table 3 *continued*

	Lemur's	*Lepi.*	Indriids	Loris's	*Adapis*	*Pelyc.*	*Noth.*	*Smilo.*
Promontory artery	sm.	sm.	sm.	sm.	sm.	sm.	sm.	sm.
Stapedial artery	lg.	vest.	lg.	vest.-0	lg.	lg.	lg.	lg.
Ascending pharyngeal artery (pharyngeo-carotid anastomisis)	0	0	0	×	0	0	0	0
Orbital frontation	br.	br.	br.	mod.-mk.	br.	br.	br.	br.
Rostral elongation	0	0	0	mod.-mk.	0	0	0	0
Premaxillary margin	undist.	undist.	undist.	dist.	undist.	undist.	undist.	undist.
Ethmoid exposure	0 (*Hapa.*)	×	occ.	×	×	?	?	?
Palatine-frontal contact	× (*Hapa.*)	0	0	0	0	?	?	?
Calcaneus	sh.	sh.	sh.	lng (lorisids, *G.crass.*)	sh.	sh.	sh.	sh.
Navicular	sh.	sh.	sh.	lng (lorisids, *G.crass.*)	sh.	sh.	sh.	?
Trochlear facet	sh.br.	sh.br.	sh.br.	nar.lng	sh.br.	sh.br.	sh.br.	?
Articular surface mandib. condyle	str.	"lip"	arc.	str.	arc.	"lip"	"lip"	str.
Paraconid	0 (*Hapa.*)	0	0	0	M_1	M_{1-3}	M_{1-2}	M_{1-2}
Paracristid	arc.	ant.	"ant."	low	ant.	obl.-med.	obl.-med.	obl.-med.
M_1 trigonid	lng	open	"open"	ex.comp.	open	comp.	comp.	comp.
Cristid obliqua M_1	buc.	buc.	mtd.	ex.med. (cheiro's)	med.	mtd	med.	mtd
M_2	buc.	buc.	med.	ex.med. (cheiro's)	med.	med.	med.	mtd
M_3	buc.	buc.	buc.	ex.med. (cheiro's)	med.	"kinks"	"kinks"	mdl-mtd
Upper incisors	red.	0	spat.	hi.cr.	spat.	red.	red.	red.
M^{1-2} shape	rect.	quad.	quad.	transv.	quad.	transv.	transv.	quad.
M^3 shape	rect.	quad.	quad.	transv.	quad.	transv.	transv.	quad.
Upper molar buccal cusps	semi-comp.	comp.	comp.	bulb.	comp.	bulb.	bulb.	semi-comp.
Upper molar protocristae	br.	br.	ex.br.	"V-U"	br.	"V"	"V"	ex.br.
Protocone M^{1-2}	br.	br.	br.	nar.	br.	nar.	nar.	br.

Key to abbreviations: Lemur's = Lemurids; *Lepi.* = *Lepilemur*; Loris's = Lorisids; *Adapis* = *Adapis*-group; *Pelyc.* = *Pelycodus*-group; *Noth.* = *Notharctus*-group; *Smilo.* = *Smilodectes*-group; cheiro's = Cheirogaleids; f. = foramen; × = present; 0 = not present; — = not applicable; fus. = fused; unfus. = unfused; post = posterior; lat. = lateral; med. = medial; cent. = central; sm. = small; lg. = large; vest. = vestigial; br. = broad; mod. = moderate; mk. = marked; dist. = distended; undist. = undistended. occ. = occasional; sh. = short; lng = long; nar. = narrow; str. = straight; arc. = arcuate; ant. = anterior; obl. = oblique; med. = medial; buc. = buccal; comp. = compressed; mtd = metaconid; red. = reduced; spat. = spatulate; hi.cr. = high crowned; rect. = rectangular; quad. = quadrate; transv. = transverse; bulb. = bulbous; "V" = "V" shaped; "U" = U shaped; ex. = extra; taxa in brackets = exceptions; *Hapa.* = *Hapalemur*; *G.crass.* = *Galago crassicaudatus*.

divergent protocristae of which the postprotocrista is less emphasized, buccolingual compression of the first three lower premolars, emphasis on the talonid and trigonid crests on the last lower premolar as well as on M_{1-3}, lower molars also distinguished by an anteriorly directed paracristid that opens the trigonid lingually, and absence of a paraconid (see Table 3). These characters of the posterior dentition point towards the conclusion that *Adapis* bears an affinity to a group of living strepsirhines which includes *Lepilemur* and the indriids, and particularly to the latter. A serious problem, of course, lurks in the respective conformations of the anterior lower dentitions. The extant forms share the almost ubiquitous strepsirhine dental comb, while *Adapis* has relatively orthal and short-crowned lower incisor teeth. If the relationships suggested by cheek-tooth morphology are accurate, it follows that *Adapis* is descended from a toothcomb-bearing ancestor. Such radical modification would not, however, be without subsequent parallel among strepsirhines: the anterior lower dentition of the subfossil indriid *Hadropithecus* shows no obvious sign of its procumbent antecedents, and an ancestor of *Daubentonia* must likewise have possessed a dental comb. Only time and further fossil discoveries will tell whether or not the lower anterior dental conformation of *Adapis* and *Leptadapis* was typical of the wider Eocene clade to which these forms belonged.

Apart from the indriid-like quadrateness of the molars of *Smilodectes*, a feature that we must conclude evolved in parallel in the two taxa, the *Pelycodus*-group shows no derived features that would link it with any particular group within Strepsirhini, including the adapids. The "lemur-like" auditory bulla of *Smilodectes, Notharctus* and *Pronycticebus* serves only to place these taxa in a sister relationship to Strepsirhini as a whole. The larger clade consisting of the *Pelycodus*-group and the strepsirhines can, however, be defined on the basis of the configuration of the auditory bulla. If, as suggested above, the relationships of *Tarsius* may lie within Strepsirhini, specifically with the lorisiform group, the spatial relationship of the ectotympanic ring to the bulla in *Tarsius* can be interpreted plausibly as a primitive retention (the ring lying within the bulla, which "captures" it in the course of development), while the tubular extension of the ectotympanic may be regarded as yet another of the uniquenesses of this highly autapomorphic primate. Figure 3 provides a very tentative hypothesis of relationships within this larger group that is consistent with the characters we have been discussing.

Fossil "tarsioids"

Numerous fossil primates, primarily distinguished by their small size, have been associated with *Tarsius* in a "tarsioid" group (e.g., Szalay, 1976, and references therein). What justification exists for this association? Historically, overall impressions from cranial material, especially of the European *Necrolemur* (Simons & Russell, 1960) and the North American *Tetonius* (Gregory, 1922), have appeared to support it: characters contributing to these impressions include relatively large orbits, V-shaped dental arcades, and tubular ectotympanics. In the postcranium, features such as elongation of the calcaneus and navicular, and in the case of the microchoerines tibio-fibular fusion, have also appeared to support this relationship, although as early as 1922 Gregory pointed out that the similarities of *Notharctus* with *Galago* were more impressive than those with *Tarsius*, and Simpson (1940, p. 196) commented on the postcranials of *Hemiacodon* that "if these bones were judged entirely on their own merits . . . probably nobody would refer them definitely to the Tarsioidea". But despite the general dissimilarities from *Tarsius* in the dentition of

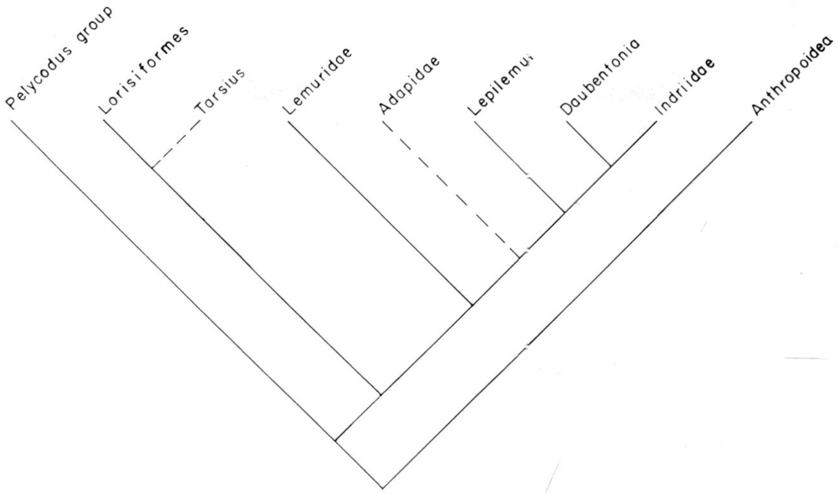

Figure 3. Tentative hypothesis of relationships among the various living and extinct primate groups.

microchoerines, anaptomorphids, and almost all omomyids, the idea that all are related in some way to the tarsier has managed to persist via the similarities in general molar morphology between *Omomys* and the extant form (e.g., Simons, 1961). These similarities, however, are mostly in the form of primitive retentions, for example the presence of a paraconid on the lower molars and presence of simple three-cusped upper molars.

The first question to be asked, then, is simply whether the fossil "tarsioids" actually constitute a monophyletic group. If all taxa that over the years have been allocated to Tarsioidea are considered together, we are unable to detect even one feature that would distinguish them as a clade. However, many of these forms (e.g., *Pseudoloris, Omomys, Tetonius, Absarokius*) are distinctive in possessing an enlarged anterior tooth, best known from preserved lower jaws. Other presumed tarsioids, in contrast, appear to have had small teeth at the front of the jaw; these forms include *Anaptomorphus, Loveina, Shoshonius, Washakius, Chumashius* and the type specimen of *Chlororhysis*. This latter "group" is distinguished by a common primitive retention of diminutive anterior teeth, and does not constitute a cohesive clade. The group with enlarged anterior teeth, which could be referred to collectively as omomyoids, also shows in M1 the presence of a prehypocone crista or its apparent derivative, a full-fledged protocone fold. This feature, as noted earlier, is distinctive of the lorisiform clade as well as of *Tarsius*. Lower molars are characterized by protoconids and metaconids melded at their bases and forming a steep wall facing on the talonid basin. This set of characters likewise distinguishes lorisiforms and *Tarsius*. In cases where postcranials are known, the omomyoid calcaneus and navicular bones are elongate to the extent typical, for example, of most *Galago* and *Microcebus*. The trochlear facet of known astragali associated with these taxa displays the derived narrow and elongate morphology otherwise seen in lorisiforms, although not in *Tarsius*. These forms are also distinguished from *Tarsius* in the diminution, if not the absence, of paraconids on at least the posterior lower molars. *Tarsius* retains large and distinct paraconids on all lower molars; lorisiforms lack paraconids altogether. Lower premolars of lorisiforms do not have truncated heels but are instead oval in cross section; this configuration also characterizes *Tarsius* and omomyoids. Relative orbital frontation and enlargement, reduced interorbital septum, and some rostral elongation also distinguish this larger group. In *Necrolemur* the ectotympanic is intrabullar; this is consistent, as we have argued above, with the other synapomorphies of the yet larger group that consists of the *Pelycodus*-group + Strepsirhini. As in the case of *Tarsius, Necrolemur* would be autapomorphic in its development of a tubular extension of the ectotympanic.

Some fossil "tarsioids" with small anterior teeth, namely *Loveina, Anaptomorphus,* and *Chlororhysis*, show similarities to *Pelycodus* in various derived features (e.g., an arcuate cristid obliqua that meets the metaconid on M_1; M_{1-2} talonid cusps poorly distinguished and incorporated in a cristid that completely encloses the talonid basin; extreme transverseness of M^{1-2}; M^2 broad lingually, with squared corners), and belong with the *Pelycodus*-group.

The affinities of *Washakius* and *Shoshonius* appear to lie with another member of the *Pelycodus*-group: *Smilodectes*. Features common to all three taxa include cristids obliquae on M_{1-2} that meet the metaconid, and a mesially directed cristid obliqua on M_3. In the upper molars, mesostyles are present on M^1. All small anterior toothed "tarsioids" possess distinct protocone folds on M^{1-2}, a character distinctive of the *Pelycodus*-group at large.

Summary and conclusions

The relationships within the strepsirhine group as reflected in Figure 1 are reasonably firmly established on the basis of morphology. One possible exception to this is the linking of *Lepilemur* and its subfossil relative *Megaladapis* with the Indriidae/Daubentoniidae clade rather than with Lemuridae, the family with which it has been traditionally associated. However, the similarities (notably in cranial and mandibular structure) which have been the primary factor in grouping *Lepilemur* with the lemurids, actually represent primitive retentions in both clades and thus do not serve to link the two.

Tarsius, of course, represents a much more difficult problem, a problem due principally to the highly autapomorphic nature of this primate. The apomorphies of the tarsier severely limit the characters potentially available for evaluation of its phylogenetic relationships, and in some cases, for example the diminution of the nasal fossa, provide false apparent synapomorphies with other primates. Our re-evaluation of the list of characters cited by Aiello (1986) and by MacPhee & Cartmill (1986) in support of a *Tarsius*-anthropoid relationship, and of other features, suggests that few provide potential evidence for this relationship. Those that do are the prebullar course of the internal carotid artery, development of an anterior accessory cavity of the auditory region, and lack of a rhinarium. In contradistinction to these apparent synapomorphies with the anthropoids, however, there exists in the tarsier a suite of characters suggesting affinity with non-anthropoids. In the widest context, it appears that the living strepsirhines, plus the adapids and the *Pelycodus*-group, are descended from a common ancestor that possessed the "lemur-like" configuration of the auditory bulla and tympanic ring. As discussed above, the specialized ear region of *Tarsius* is also plausibly derived from this condition. Within this larger group all extant taxa, including *Tarsius*, are united by the presence of a grooming claw on the second pedal digit. Yet more specifically, *Tarsius* and the lorisiforms exhibit a prehypocone crista on M^{1-2}, ovoid lower premolars, lower molar protoconids and metaconids melded at their bases and forming a steep wall facing on the talonid, buccal expansion of an angular hypoconid on M_{1-2}, downward extension of the alveolar margin of the premaxilla, some anterior rostral elongation, and relative orbital enlargement and frontation. Other features that appear to be suggestive if not demonstrative of a tarsier–lorisiform link include: loss of the stapedial artery in adults of this group, elongation of the calcaneus and navicular (except in *Galago crassicaudatus* and in lorisids), and a high degree of chromosome homology (at least between the tarsier and *G. crassicaudatus*).

Characters may thus be discerned in the tarsier that point towards both anthropoid and non-anthropoid affinities. The fact that apparent synapomorphies exist at different levels within the evolutionary hierarchy of the non-anthropoid clade (for which Prosimii would be the obvious name) may well suggest that a hypothesis emphasizing these is more robust than one suggesting a tarsier–anthropoid clade. Clearly, at the current state of our knowledge the inconveniently autapomorphic *Tarsius* continues to elude definitive phylogenetic placement; but nesting this form with the lorisiforms is a possibility that deserves serious consideration.

Acknowledgements

We thank the Richard Lounsbery Foundation for its support, and Fred Grine and Lawrence Martin for their invitation to participate in this symposium. Peter Andrews, Fred Grine, and several anonymous reviewers kindly commented on an earlier draft of the manuscript. This is Contribution no. 20 of the Lounsbery Laboratory of Biological Anthropology, American Museum of Natural History.

References

Aiello, L. (1986). The relationships of the Tarsiiformes: a review of the case for the Haplorhini. In (B. Wood, L. Martin & P. Andrews, Eds) *Major Topics in Primate and Human Evolution*, pp. 47–65. Cambridge: Cambridge University Press.

Baba, M., Weiss, M. L., Goodman, M. & Czelusniak, K. (1982). The case of tarsier hemoglobin. *Syst. Zool.* **31,** 156–165.

Bugge, J. (1974). The cephalic arterial system in insectivores, primates, rodents, and lagomorphs, with special reference to systematic classification. *Acta Anat.* **87** (Suppl. 62), 1–60.

Cartmill, M. (1972). Arboreal adaptations and the origin of the order of the Primates. In (R. Tuttle, Ed.) *Functional and Evolutionary Biology of Primates*, pp. 97–122. Chicago: Aldine-Atherton.

Cartmill, M. (1975). Strepsirhine basicranial structures and the affinities of the Cheirogaleidae. In (W. P. Luckett & F. Szalay, Eds) *Phylogeny of the Primates: a Multidisciplinary Approach*, pp. 313–354. New York: Plenum Press.

Cartmill, M. (1978). The orbital mosiac in prosimians and the use of variable traits in systematics. *Folia primat.* **30,** 89–114.

Cartmill, M. (1981). Morphology, function and evolution of the Anthropoid postorbital septum. In (R. Ciochon & A. Chiarelli, Eds) *Evolutionary Biology of the New World Monkeys and Continental Drift*, pp. 243–274. New York: Academic Press.

Castenholz, A. (1965). Uber die struktur der Netzhautmitte bei Primaten. *Z. Zellforsch* **65,** 646–661.

Castenholz, A. (1984). The eye of *Tarsius*. In (C. Niemitz, Ed.) *Biology of Tarsiers*, pp. 303–318. Stuttgart: Gustav Fischer Verlag.

Cave, A. J. E. (1973). The primate nasal fossa. *Biol. J. Linn. Soc.* **5,** 377–387.

Dene, H., Goodman, M. & Prychodko, W. (1976). Immunodiffusion evidence on the phylogeny of the primates. In (M. Goodman, E. Tashian & J. Tashian, Eds) *Molecular Anthropology*, pp. 171–195. New York: Plenum.

Ford, S. (1980). A systematic revision of the Platyrrhini based on features of the postcranium. Ph.D. thesis, University of Pittsburgh.

Gingerich, P. D. (1979). Phylogeny of Middle Eocene Adapidae (Mammalia, Primates) in North America: *Smilodectes* and *Notharctus*. *J. Paleont.* **53,** 153–163.

Gregory, W. K. (1922). *The Origin and Evolution of the Human Dentition*. Baltimore: Williams and Wilkins.

Hershkovitz, P. (1977). *Living New World Monkeys (Platyrrhini)*, Vol. 1. Chicago: University of Chicago Press.

Hill, W. C. O. (1955). *Primates, Comparative Anatomy and Taxonomy*. vol. II, *Haplorhini, Tarsioidea*. Edinburgh: Edinburgh University Press.

Hofer, H. (1977). The anatomical relations of the ductus vomeronasalis and the occurrence of taste buds in the papilla palatina of *Nycticebus coucang* (Primates, Prosimiae), with remarks on strepsirhinism. *Morph. Jahrb.* **123,** 836–856.

Hofer, H. O. (1980). The external anatomy of the oro-nasal region of primates. *Morph. Anthrop.* **71,** 232–249.

Jouffroy, F. K., Berge, C. & Niemitz, C. (1984). Comparative study of the lower extremity in the genus *Tarsius.* In (C. Niemitz, Ed.) *Biology of Tarsiers*, pp. 167–190. Stuttgart: Gustav Fischer Verlag.

Koenigswald, G. H. R. von (1979). Ein Lemurenrest aus dem eozänen Ölschiefer der Grube Messel bei Darmstadt. *Paläont. Zeitschr.* **53,** 63–76.

Kollmer, W. (1930). Zur Kenntnis des Auges der Primaten. *Anat. Entwickl.* **93,** 679–722.

Leutenegger, W. (1973). Maternal fetal weight relationships in primates. *Folia primat.* **20,** 280–293.

Luckett, W. P. (1974). Comparative development and evolution of the placenta in primates. *Contr. Primat.* **3,** 142–234.

Luckett, W. P. (1975). Ontogeny of the fetal membranes and the placenta: their bearing on primate phylogeny. In (W. Luckett & F. Szalay, Eds) *Phylogeny of the Primates: a Multidisciplinary Approach*, pp. 157–182. New York: Plenum.

Luckett, W. P. (1976). Cladistic relationships among primate higher categories: evidence of the fetal membranes and placenta. *Folia primat.* **25,** 245–276.

MacPhee, R. D. E. & Cartmill M. (1986). Basicranial structures and primate systematics. In (D. Swindler & J. Erwin, Eds) *Comparative Primate Biology* Vol. 1, *Systematics, Evolution and Anatomy*, pp. 219–275. New York: Alan R. Liss.

Musser, G. G. & Dagosto, M. (1987). The identity of *Tarsius pumilus*, a pygmy species endemic to the montane mossy forests of central Sulawesi. *Am. Mus. Novit.* **2867,** 1–53.

Pocock, R. I. (1918). On the external characters of the lemurs and *Tarsius. Proc. zool. Soc. Lond.* for 1918, 19–53.

Poorman, P. A., Cartmill, M., MacPhee, R. & Moses, M. (1985). The banded karyotype of *Tarsius bancanus* and its implications for primate phylogeny. *Am. J. phys. Anthrop.* **66,** 215.

Rohen, J. W. (1966). Zur Histologie des Tarsiusauges. *Alb. v. Graefes Arch. Klin. Exp. Ophthal.* **269,** 299–317.

Saban, R. (1963). Contribution a l'étude de l'os temporal des primates. *Mem. Mus. natl. Hist. nat., ser. A, Zoologie* **29,** 1–378.

Schwartz, J. H. (1984). What is a tarsier? In (N. Eldredge & S. Stanley, Eds) *Living Fossils*, pp. 38–49. New York: Springer Verlag.

Schwartz, J. H. (1986). Primate systematics and a classification of the Order. In (D. Swindler, Ed.) *Comparative Primate Biology* Vol. 1, pp. 1–41. New York: Alan R. Liss.

Schwartz, J. H. & I. Tattersall (1985). Evolutionary relationships of living lemurs and lorises (Mammalia, Primates) and their potential affinities with European Eocene Adapidae. *Anthrop. Pap. Am. Mus. nat. Hist.* **60,** 1–100.

Simons, E. L. (1961). Notes on Eocene tarsioids and a revision of some Necrolemurinae. *Bull. Br. Mus.* (nat. Hist.), *Geol. Ser.* **5,** 45–69.

Simons, E. L. & Russell D. E. (1960). Notes on the cranial anatomy of *Necrolemur. Breviora Mus. comp. Zool.* **127,** 1–14.

Simpson, G. G. (1940). Studies on the earliest primates. *Bull. Amer. Mus. nat. Hist.* **77,** 185–212.

Starck, D. (1975). The development of the chondrocranium in primates. In (W. Luckett & F. Szalay, Eds) *Phylogeny of the Primates: a Multidisciplinary Approach*, pp. 127–155. New York: Academic Press.

Starck, D. (1984). The nasal cavity and nasal skeleton of *Tarsius.* In (C. Niemitz, Ed.) *Biology of Tarsiers*, pp. 275–290. Stuttgart: Gustav Fischer Verlag.

Stephan, H. (1984). Morphology of the brain in *Tarsius.* In (C. Niemitz, Ed.) *Biology of Tarsiers*, pp. 319–344. Stuttgart: Gustav Fischer Verlag.

Szalay, F. S. (1975). Phylogeny of primate higher taxa: the basicranial evidence. In (W. Luckett & F. Szalay, Eds) *Phylogeny of the Primates: a Multidisciplinary Approach*, pp. 91–125.

Szalay, F. S. (1976). Systematics of the Omomyidae (Tarsiiformes, Primates): taxonomy, phylogeny, adaptations. *Bull. Am. Mus. nat. Hist.*, **156,** 157–450.

Szalay, F. S. (1977). Constructing primate phylogenies: a search for testable hypotheses with maximal empirical content. *J. hum. Evol.* **6,** 3–18.

Tattersall, I. & Schwartz, J. H. (1974). Craniodental morphology and the systematics of the Malagasy lemurs (Primates, Prosimii). *Anthrop. Pap. Am. Mus. nat. Hist.* **52,** 141–192.

Wolin, L. & Massopust, L. (1970). Morphology of the primate retina. In (C. Noback & W. Montagna, Eds) *The Primate Brain*, pp. 1–27. New York: Appleton-Century-Crofts.

Woollard, H. H. (1926). Notes on the retina and lateral geniculate body in *Tupaia, Tarsius, Nycticebus* and *Hapale. Brain* **49,** 77–104.

Terry Harrison

Department of Anthropology, New York University, 25 Waverly Place, New York, NY 10003, U.S.A.

Received 27 March 1987
Revision received 25 June 1987
and accepted 30 June 1987
Publication date November 1987

Keywords: phylogenetic relationships, Oligocene, Miocene, anthropoids, catarrhines, parapithecids, *Oligopithecus, Propliopithecus, Pliopithecus, Dendropithecus, Proconsul, Victoriapithecus.*

The phylogenetic relationships of the early catarrhine primates: a review of the current evidence

This paper presents a review of the evolutionary relationships of the early catarrhine primates. The first stage of the analysis involves the reconstruction of the inferred ancestral morphotypes of the major groups of extant anthropoids. The introduction of the fossil taxa into the phylogenetic scheme represents the second and final stage of the analysis. The results of this cladistic analysis suggests that: (1) the parapithecids are a specialized group of basal anthropoids, (2) *Oligopithecus savagei* may represent the earliest recognizable catarrhine, (3) *Propliopithecus* (= *Aegyptopithecus*) and *Pliopithecus* apparently represent the successive sister taxa to the modern catarrhines, (4) *Dendropithecus* and *Proconsul* are best regarded as basal catarrhines of modern aspect, and (5) *Victoriapithecus* is a primitive cercopithecoid monkey which represents the sister taxon of the extant Old World monkeys.

Journal of Human Evolution (1987) **16,** 41–80

Introduction

The wealth of new fossil evidence of early catarrhine primates from the Oligocene deposits of the Fayum in Egypt and the early Miocene sites of East Africa collected over the past decade, as well as the increasing breadth and quality of the fossil samples from later Miocene sites in Eurasia, has led to a reformulation of our previous ideas concerning the evolutionary relationships of the catarrhines. Although detailed analyses of much of this new fossil material have yet to be published, it is possible to present here a preliminary synthesis of the available evidence to provide an up-dated phylogenetic scheme. The aim of this paper, therefore, is to review the phylogenetic relationships of the early catarrhine primates. It is not intended to be a systematic survey of the relationships of all known taxa, many of which are too poorly preserved to be able to realistically assess their affinities. There continues to be considerable debate over fossil primates that are relatively well-represented (e.g., *Aegyptopithecus, Oreopithecus* and *Proconsul*); thus it seems rather academic, or perhaps even fruitless, to speculate about the evolutionary relationships of a taxon based on a single isolated tooth or a fragmentary mandible. For this reason, the present review will concentrate on taxa that are relatively well-known, as these presumably offer the greatest potential for successfully determining their phyletic affinities. The primary objective is to review the initial radiation of the catarrhines in order to assess the relationships of the early fossil forms with the living anthropoid primates.

The first stage in this analysis is to reconstruct the inferred ancestral morphotypes of the major groups of extant anthropoids based on neontological data. The Recent taxa are clearly the most important basis for constructing the initial phyletic framework because they permit a more comprehensive documentation of the anatomy, which may in turn be supported by behavioral, ecological and molecular studies. The neontological evidence thus provides the most secure and most easily verifiable set of inferred relationships between taxa. Moreover, the exclusion of fossil anthropoids at this stage of the analysis avoids the potential pitfalls (and also the temptations) of circularity in determining morphocline polarities based on the relative temporal position of taxa. This approach is obviously practical only when the extant groups exhibit some diversity. Fortunately, the

0047–2484/87/010041 + 40 $03.00/0

Anthropoidea is represented today by a diverse array of living forms, and this greatly facilitates the reconstruction of the nest of ancestral morphotypes within the group.

Based on neontological studies, it is possible to construct a generally accepted set of relationships among the major extant groups within the Anthropoidea (Figure 1). For each of these groups, the ancestral morphotype can be defined as a series of individual morphological characters. Using only the key osteological and dental features from each of these morphotypes (Table 1), it is then possible to assess the phyletic affinities of the fossil taxa relative to the extant anthropoid primates. The introduction of the fossil catarrhines into the phylogenetic framework in Figure 1 will thus represent the second and final stage of the analysis. It is important to note that, although the fossils are introduced at a later stage than the living taxa, they are not considered as a special category because of their potential to add a temporal perspective. It is simply that they are taxa about which we know very much less, and therefore inferences concerning their relationships are likely to be less well supported and less easily testable.

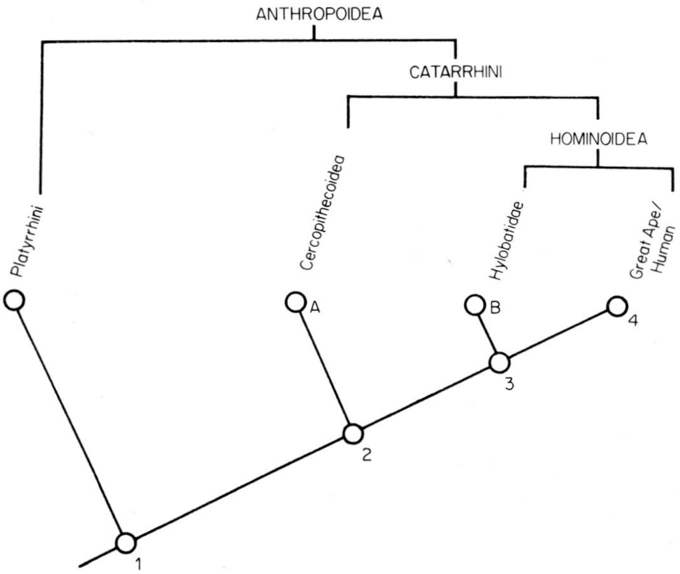

Figure 1. A cladogram illustrating the phylogenetic relationships and the taxonomy of the modern anthropoid primates. The characters defining Nodes 1, 2, 3 and A are presented in Table 1.

Reconstruction of the inferred ancestral morphotypes

Based on comparisons of extant anthropoid primates, it is possible to reconstruct inferred ancestral morphotypes for the major groups of primates within the Anthropoidea (Table 1). In order to assess the phylogenetic relationships of the early Old World anthropoids, it is necessary to define the ancestral morphotypes for the Anthropoidea, the Catarrhini, the Hominoidea and the Cercopithecoidea. Owing to the nature of the fossil evidence, only the dental and skeletal characters that define each morphotype are discussed.

Table 1 **A list of skeletal and dental characters defining the principal nodes in Figure 1**

Node 1: Ancestral anthropoid morphotype

1. Face small in relation to the size of the neurocranium.
2. Premaxilla relatively small.
3. Nasal process of the premaxilla makes slight contact with the nasal bone.
4. Subnasal region of the premaxilla relatively low.
5. Incomplete closure of the interpremaxillary suture.
6. Nasal aperture ovoid in shape, higher than broad with rounded or tapering inferior margin.
7. Greatest breadth of nasal aperture about midway up.
8. Infraorbital foramina ($N = 1$–3) situated close to the inferior margin of the orbit.
9. Orbits completely separated from the temporal fossa by a bony septum.
10. Inferior orbital fissure relatively large; lateral orbital fissure small.
11. Ethmoid enters into the orbital mosaic.
12. Lacrimal bone completely or almost completely contained within the orbit.
13. Anterior margin of the orbit overlaps with the superior portion of the nasal aperture.
14. Anterior margin of the orbit situated above C-P2.
15. Relatively large maxillary and sphenoidal sinuses.
16. Turbinals of the nasal fossa reduced in number and complexity.
17. Anterior root of zygomatic arch situated above M1–M2 and close to the alveolar margin of the maxilla.
18. Well-developed postglenoid foramen.
19. Stapedial artery absent; promontory artery enlarged.
20. Annular ectotympanic connected to the squamosal portion of the temporal.
21. Subarcuate fossa of the temporal deep and well-developed.
22. No development of supraorbital brow ridges.
23. Fusion of metopic suture of frontal bone occurs early in life.
24. Palate relatively shallow and long and narrow with anteriorly converging toothrows.
25. Incisive foramen large, elliptical and paired. Situated close to the alveolar margin of the incisors.
26. Palatine extends anteriorly as far as M^1.
27. Mandibular symphysis fused early in life.
28. Mandibular symphysis with superior transverse torus only.
29. Mandible moderately high with a constant depth below P2–M3.
30. Mandibular toothrows converge anteriorly.
31. Angular process of mandible reduced.
32. Coronoid process of mandible higher than the condylar process.
33. Dental formula 2.1.3.3., with loss of upper and lower P1.
34. Upper I1 spatulate and slightly higher than broad.
35. Upper I2 high, narrow and caniniform.
36. Upper I2 slightly smaller than I1.
37. Upper canine high-crowned and projecting with slight bilateral compression and a single narrow mesial groove not extending onto the root of the tooth.
38. Upper premolars broad and triangular to ovoid with the buccal cusp only slightly higher than the lingual cusp.
39. Upper molars with the following morphological features:
    Short and broad, narrowing lingually. Length–breadth index of M^2: <80
    Trigon cusps high and conical.
    Crista obliqua well-developed.
    Metaconule distinct.

Table 1—*continued*

Node 1—continued

Lingual crest of paracone does not meet the mesiobuccal crest of the protocone.
Mesial fovea indistinct.
Narrow lingual cingulum around the lingual margin of the protocone.
Hypocone extremely small or absent.
No connection between the hypocone and the trigon.
Size differential: M3 < M1 < M2.

40. Lower incisors low-crowned and moderately broad.
41. Lower canine moderately high crowned and moderately bilaterally compressed.
42. Upper and lower canines sexually dimorphic.
43. Lower P3 single-cusped, moderately high-crowned, broad and ovoid, with no extension of enamel onto the mesial root of the tooth. No contact with the upper canine.
44. Lower P4 bicuspid, ovoid in outline and broader than long, with subequal cusps and large foveae.
45. Lower molars with the following morphological features:
    Rectangular to ovoid in shape, much longer than broad.
    Crests and cusps high and sharp.
    Mesial fovea large and well-defined.
    Hypoconulid very small and lingually placed.
    Paraconid vestigial.
    Cristid obliqua long and very oblique.
    Distal fovea not developed.
    Narrow but well-developed buccal cingulum.
    Enamel layer thin and free from secondary folds and wrinkling.
    Size differential: M3 < M2 < M1.
46. Lower M3 erupts after all of the lower deciduous premolars have been replaced.
47. Incisors small relative to the size of the molars.
48. Upper dP3 bicuspid and triangular with no transverse crest linking the cusps. Buccal cusp much higher than the lingual cusp.
49. Upper dP4 with hypocone vestigial or absent.
50. Lower dP3 long and ovoid with a single main cusp.
51. Lower dP4 with a reduced paraconid. Trigonid shortened to form a well-defined mesial fovea. Hypoconulid absent or very small and lingually placed.
52. Relatively short cervical vertebral region.
53. Vertebral centra relatively long.
54. Thoracic vertebrae do not protrude into the thoracic cavity.
55. Number of thoracic vertebrae: 13.
56. Number of lumbar vertebrae: 6–7.
57. Number of sacral vertebrae: 3.
58. Number of caudal vertebrae: 20–30.
59. Tail relatively long.
60. Vertebral border of the scapula moderately long.
61. Scapula laterally placed.
62. Scapula with well-developed acromion and coracoid processes.
63. Clavicle short and stout.
64. Sternum long and slender with 6–7 sternal elements.
65. Angle between corpus and collum of ribs greater than 90 degrees.
66. Thorax medio-laterally narrow.
67. Humerus with the following morphological features:
    Lateral trochlear keel absent or poorly-developed.
    Breadth of the capitulum on average equal to or exceeding that of the trochlea.
    Entepicondylar foramen present.
    Medial epicondyle strongly projecting and medially directed.

Table 1—*continued*

Node 1—continued

Dorsal epitrochlear fossa well-developed.
Radial and coronoid fossa shallow.
Olecranon fossa shallow and poorly defined.
Mean length-breadth index of the distal extremity: 35–45.
Slight sinusoidal curvature of the shaft.
Shaft moderately short and stout.
Slight torsion of the humeral head.
Head ovoid and asymmetrical.
Humeral head only slightly larger than the femoral head.

68. Ulna with the following morphological features:
Relatively very high olecranon process.
Trochlear notch relatively high and narrow.
Trochlear keel absent or slightly developed.
Coronoid process moderately projecting and vertically in line with the olecranon beak.
Shaft slightly recurved anteriorly.
Shaft strongly bilaterally compressed.
Styloid process large and well-developed.
Direct contact between the styloid process and the triquetrum and the pisiform.
No development of an intra-articular meniscus.
Restricted distal radio-ulnar articulation.
Distal ulna antero-posteriorly long and medio-laterally narrow.
69. Restricted range of extension at the humero-ulna articulation.
70. Radial head elliptical in outline.
71. Styloid process of radius well-developed.
72. Radius relatively short and stout.
73. Lunate relatively long and narrow.
74. Scaphoid lacks a distally directed beak-like process.
75. Os centrale free.
76. Capitate proximo-distally relatively long, with a medio-laterally narrow head.
77. Carpo-metacarpal joint of the thumb of "modified hinge" type.
78. Shallow carpal tunnel.
79. Distal extremity of the metapodial heart-shaped and narrow dorsally.
80. Prehallux present as a component of the tarso-metatarsal joint.
81. Pollex long and well-developed.
82. Phalanges moderately long and slightly recurved.
83. Digital projection formula: 3>4>2>5>1.
84. Calcaneus long and narrow with facets of primitive shape and position.
85. Lateral and medial condyles of the proximal tibia relatively symmetrical.
86. Femur with the following morphological features:
Head continuous with the neck posteriorly.
Femur head relatively small.
Neck relatively long with a high neck angle.
Tubercle present on the neck of the femur.
Shallow but extensive trochanteric fossa.
Well-developed third trochanter on shaft.
Shaft straight and relatively gracile.
Medial and lateral condyles relatively symmetrical.
Long axis of shaft almost vertically aligned.
Distal extremity medio-laterally broad and anteroposteriorly short.
Patella groove high and narrow.
87. Long and relatively narrow patella.
88. Forelimb generally shorter than hindlimb.
89. Relatively long and narrow ilium with convex iliac fossa.

Table 1—*continued*

Node 2: Ancestral catarrhine morphotype.

5. Complete closure of interpremaxillary suture early in life.
10. Lateral orbital fissure absent.
13. Anterior margin of the orbit situated above P3–P4.
18. Postglenoid foramen very reduced.
20. Tubular ectotympanic.
21. Subarcuate fossa of temporal well-developed, but shallow.
28. Mandibular symphysis with superior transverse torus only or dominant over inferior transverse torus.
29. Mandible shallows posteriorly.
30. Mandibular toothrows converge only slightly anteriorly.
31. No development of an angular process on the mandible.
33. Dental formula 2.1.2.3., with loss of upper and lower P2.
38. Upper premolars with the buccal cusp much higher than the lingual cusp.
39. Upper molars with the following morphological features:
    Crown rectangular.
    Length-breadth index of M2: < 90.
    Trigon cusps low and conical.
    Metaconule indistinct.
    Lingual crest of the paracone meets the mesio-buccal crest of the protocone.
    Well-defined mesial fovea.
    Hypocone large and well-developed.
    Hypocone connected to the trigon by a short crest.
40. Upper and lower incisors moderately high-crowned.
43. Lower P3 single-cusped, moderately high-crowned and long and quite narrow with slight to moderate extension of enamel onto the mesial root. Moderately long honing face for contact with the upper canine.
44. Lower P4 as long as broad or slightly longer than broad.
45. Lower molars with the following morphological features:
    Crown slightly longer than broad.
    Cusps moderately high and conical.
    Crests low and rounded.
    Mesial fovea well-defined, but short and broad.
    Hypoconulid well-defined and medially or buccally placed.
    Paraconid absent.
    Cristid obliqua relatively short and only slightly oblique.
    Distal fovea well-defined.
    Size differential: M1 < M2 < M3.
48. Upper dP3 with well developed lingual cusp.
49. Upper dP4 rectangular with small but well-developed hypocone; anterior transverse crest well-developed and delimiting a true mesial fovea; cusps quite low and rounded.
51. Lower dP4 with small, but well-developed hypoconulid placed medially or buccally; distal fovea small but distinct; paraconid very small or absent.
67. Entepicondylar foramen of distal humerus absent.
77. Carpo-metacarpal joint of the thumb of "saddle" type.
86. Third trochanter of femur rudimentary or absent.

Node A: Ancestral cercopithecoid morphotype.

6. Nasal aperture high and narrow with V-shaped inferior margin.
8. Increased number of infraorbital foramina.
9. Secondarily narrowed interorbital region.
10. Inferior orbital fissure greatly reduced.
15. Maxillary sinus greatly restricted.
28. Mandibular symphysis with dominant inferior transverse torus.
30. Mandibular toothrows almost parallel.

Table 1—*continued*

Node A—continued

37. Upper canines high-crowned, strongly bilaterally compressed with a single mesial groove extending onto the cervix of the root in males. Strongly sexually dimorphic.
38. Upper premolars relatively long and D-shaped.
39. Upper and lower molars high-crowned, long and narrow and modified to form a bilophodont pattern. No crista obliqua.
40. Incisors relatively narrow and very high-crowned and slightly procumbent.
41. Lower canines high-crowned and slender. Strongly sexually dimorphic.
43. Lower P3 long and narrow with a marked extension of enamel onto the mesial root of the tooth. Highly sectorial.
44. Lower P4 longer than broad.
45. Loss of hypoconulid on lower M1 and M2.
48. Upper and lower deciduous premolars very long, rectangular and bilophodont.
67. Humerus with the following morphological features:
    Medial epicondyle slightly posteriorly directed and abbreviated.
    Olecranon fossa relatively deep and well-defined.
    Mean length–breadth index of distal extremity >50.
    Shaft with marked sinusoidal curvature and pronounced muscle markings.
74. Scaphoid specialized with a well-developed distally directed beak-like process.
80. Prehallux lost?
82. Phalanges relatively short and quite stout.
86. Femur with the following morphological features:
    Head is continuous with the neck superiorly as well as posteriorly.
    Neck angle of femur is relatively low.

Node 3: ancestral hominoid morphotype.

24. Palate relatively broad anteriorly.
32. Condylar process of mandible equal in height or higher than the coronoid process.
34. Upper I1 low-crowned and broad.
38. Upper molars with the following morphological features:
    Moderately short and broad.
    Length–breadth index of M2: > 85.
    Trigon cusps low and quite rounded.
    Reduced lingual cingulum.
40. Lower incisors low-crowned and broad.
43. Lower P3 with moderately short honing face. Reduced sectoriality.
45. Lower molars broad with reduced buccal cingulum. Reduced M_3.
47. Incisors large relative to the size of the molars.
52. Increased length of the cervical vertebral region.
53. Increased diameter and reduced length of the vertebral centra.
54. Thoracic vertebrae protrude into the thoracic cavity.
56. Number of lumbar vertebrae: 5.
57. Number of sacral vertebrae: 4–5.
58. Number of caudal vertebrae: 6.
59. No external tail.
60. Scapula with elongated vertebral border.
61. Scapula dorsally positioned and redirection of the glenoid cavity.
62. Large acromion and coracoid processes on the scapula.
63. Clavicle relatively elongated.
64. Sternum short and broad with reduced number of elements.
65. Angle between the corpus and the collum of the ribs less than 90 degrees.
66. Thorax broad.

Table 1—*continued*

Node 3—continued

67. Humerus with the following morphological features:
    Medial trochlear keel more distally projecting relative to the medial epicondyle.
    Olecranon fossa relatively deep and well-defined.
    Modified curvature of the shaft.
    Marked torsion of the humeral head.
    Humerus head large, symmetrical and rounded.
    Humeral head much larger than the femoral head.
68. Ulna with the following morphological features:
    Very reduced olecranon process.
    Trochlear notch relatively low and broad.
    Shaft more robust and less bilaterally compressed.
    Distal ulna antero-posteriorly short and broad.
    Development of an intra-articular meniscus which partially isolates the styloid process from the carpus.
    Extensive distal radio-ulna articulation.
69. Greater range of extension at the humero–ulna joint.
70. Head of radius circular.
71. Styloid process of radius greatly reduced.
73. Lunate relatively broad due to the expansion of the radial articular facet.
76. Capitate proximo-distally relatively short.
78. Carpal tunnel deep for powerful long digital flexors.
79. Distal extremity of metacarpals quadrilateral.
82. Phalanges long and recurved.
84. Calcaneus short and broad with modified talo-calcaneal facets
85. Lateral and medial condyles of the proximal tibia asymmetrical.
86. Femur with the following morphological features:
    Head clearly distinct from the neck.
    Relatively large femur head.
    Tubercle on femur neck absent.
    Deep and restricted trochanteric fossa.
    Medial epicondyle much larger than the lateral epicondyle.
    Long axis of the shaft tilted laterally.
    Patella groove low and broad.
87. Patella broad and ovoid.
88. Hindlimbs shorter than the forelimbs.
89. Ilium broad with concave iliac fossa.

Sources: Harrison (1982, 1987) and references therein.

The ancestral anthropoid morphotype

The inferred ancestral morphotype of the Anthropoidea is represented by Node 1 of the cladogram in Figure 1, and the individual characters defining the morphotype are listed in Table 1. It is important to note that because the ancestral anthropoid morphotype is represented on the cladogram as the basal node, many of the included characters are retentions from the more primitive primate morphotypes, and do not necessarily represent characters defining the anthropoid clade. Primitive features have been included in the morphotype only if they have undergone subsequent transformation in the ancestral morphotypes of the subgroups under discussion here. Nevertheless, a large number of important derived characters, including the detailed structure of the facial skeleton and ear region, and modifications of the dentition and mandibular morphology confirm that the Recent Anthropoidea is a monophyletic group.

The major derived features of the inferred ancestral anthropoid morphotype that are most useful for assessing the relationships of the fossil forms are as follows: (1) the incisive region of the premaxilla is slightly broadened (Cachel, 1979; Harrison, 1982); (2) the facial region is abbreviated, and is relatively small when compared with the size of the enlarged neurocranium (Vogel, 1966, 1968; Delson & Andrews, 1975); (3) the metopic suture of the frontal bone is fused very early in life (Le Gros Clark, 1959; Simons, 1972); (4) the infraorbital foramen is situated close to the inferior margin of the orbit (this feature is also present in *Tarsius*, and may, therefore, represent a primitive haplorhine trait) (Hershkovitz, 1977; Harrison, 1982); (5) the annular ectotympanic is fused to the squamosal portion of the temporal (Le Gros Clark, 1959; Szalay, 1975*a*; Hershkovitz, 1977; Szalay & Delson, 1979); (6) the stapedial artery is usually absent and the promontory artery is enlarged (Szalay, 1975*a*; Archibald, 1977; Szalay & Delson, 1979) (7) the orbits are separated from the temporal fossa by a relatively complete postorbital septum, which is perforated only by a large inferior orbital fissure and a small lateral orbital fissure (Le Gros Clark, 1959; Hershkovitz, 1977); (8) the orbits are almost fully frontally directed (Le Gros Clark, 1959); (9) the lacrimal bone is completely or almost completely contained within the orbit (Hill, 1957; Le Gros Clark, 1959); (10) the mandibular symphysis is fused (Hill, 1957; Simons, 1972; Hershkovitz, 1977); (11) the mandibular corpus is moderately high, with a reduced angular process (Avis, 1961; Hershkovitz, 1977); (12) the P_4 is bicuspid, with subequal cusps and large, well-defined mesial and distal foveae; (13) the M_3 erupts after all of the lower deciduous premolars have been replaced (Kay & Simons, 1983); (14) the carpo-metacarpal joint of the pollex is of a "modified hinge" type (Napier, 1961, 1962); and (15) a large dorsal epitrochlear fossa is present on the posterior aspect of the medial epicondyle of the distal humerus (Conroy, 1976; Harrison, 1982).

Ancestral catarrhine morphotype

Reconstruction of the ancestral catarrhine morphotype based purely on neontological evidence is somewhat problematic. The major difficulty is that the two extant superfamilies of the Catarrhini, the Cercopithecoidea and the Hominoidea, both appear to have developed derived and structurally divergent cranial and postcranial patterns relative to the primitive catarrhine condition. Although the large number of uniquely derived morphological features enables the cercopithecoids and hominoids to be recognized as distinct and easily definable clades, the precise identification of characters defining their common ancestral morphotype is much more difficult. It presents us with the onerous task of trying to infer the primitive conditions for a whole series of characters for which the living forms exhibit only the extreme poles of divergent transformation sequences. Under these conditions it is uncertain whether the primitive pattern is more cercopithecoid-like, more ape-like, intermediate or more like the condition seen in the nearest living out-group, the New World monkeys. It is certainly tempting, when viewing the primates as a *scala naturae*, to infer that the cercopithecoid-like pattern is primitive, as has been assumed by some workers in the past (e.g., Le Gros Clark, 1937, 1959; Corruccini *et al.*, 1976; Pickford, 1982). However, this does not appear to be a valid or useful criterion for determining the polarity of transformation in many of the character states in extant catarrhines (see Fleagle, 1983*a*). Nonetheless, despite the difficulties involved, a broadly defined ancestral catarrhine morphotype can be constructed, and is presented here as Node 2 of the cladogram in Figure 1 and Table 1. It is interesting and rather reassuring to note that the inferred ancestral morphotype presented here is similar in most respects to that proposed by Delson & Andrews (1975) and Szalay & Delson (1979).

The most important dental and skeletal features defining the ancestral catarrhine morphotype are as follows: (1) the development of a tubular ectotympanic, (2) the postglenoid foramen is very reduced, (3) the lateral orbital fissure is absent, (4) the dental formula is reduced to 2.1.2.3, with the loss of the upper and lower P2, (5) the lower P3 has been modified into a sectorial tooth for occlusion with the upper canine, (6) a modified molar morphology (for details see Table 1, Node 2), (7) the entepicondylar foramen of the distal humerus is absent, and (8) the carpo-metacarpal joint of the thumb is of a "saddle" type (Le Gros Clark, 1959; Napier, 1961, 1962; Hill, 1966; Simons, 1972; Delson & Andrews, 1975; Hershkovitz, 1977; Szalay & Delson, 1979; Harrison, 1982).

Ancestral cercopithecoid morphotype

The extant Old World monkeys, the Cercopithecoidea, are characterized by a number of shared derived features of the dentition, cranium and postcranium which distinguish them from the inferred primitive catarrhine morphotype. The ancestral cercopithecoid morphotype, represented by Node A of the cladogram in Figure 1, has been well-established, at least on the basis of cranio-dental features, by Delson (1975), Szalay & Delson (1979) and Strasser & Delson (1987), and need not be extensively elaborated on here. However, several additional cranial characters are included, which further support the view that the extant Old World monkeys represent a well-defined and relatively specialized clade.

As suggested by Vogel (1966, 1968), the cranium of colobine monkeys probably conforms quite closely to the primitive catarrhine condition. However, the development of a high and narrow nasal aperture, a reduced inferior orbital fissure, a secondarily narrowed interorbital region, increased multiplicity of the infraorbital foramina, a relatively restricted maxillary sinus, almost parallel mandibular tooth rows, and a mandibular symphysis with a robust inferior transverse torus or a "simian shelf" are all derived features shared by other cercopithecoid monkeys.

The postcranial morphology of the colobine monkeys also appears to conform in many respects to the primitive catarrhine morphotype. However, it is important to dispell the widespread and longstanding notion that Old World monkeys retain an essentially conservative catarrhine postcranial pattern relative to which the hominoid condition is derived (e.g., Le Gros Clark, 1959; Delson & Andrews, 1975; Corruccini *et al.*, 1976; Andrews, 1981; Pickford, 1982; Temerin & Cant, 1983). The cercopithecids clearly share a cluster of important structural modifications of the elbow, wrist, ankle and hip joints, which presumably represent autapomorphies (Napier & Davis, 1959; Jolly, 1967, 1972; O'Connor, 1975, 1976; Szalay, 1975*b*; Szalay & Delson, 1979; Fleagle, 1983*a*; Rose, 1983; Strasser & Delson, 1987) (See Table 1, Node A). These specializations of the postcranium are largely associated with increasing the effectiveness of the movements of the major joints of the limbs in the sagittal plane (Harrison, 1982; Rose, 1983).

Ancestral hominoid morphotype

The ancestral hominoid morphotype, represented by Node 3 of the cladogram in Figure 1, is characterized by a large number of distinctive morphological features. Although several dental and cranial characters are tentatively included, the ancestral hominoid morphotype is defined largely on the basis of postcranial specializations.

Primitively, at least, the hominoids appear to retain the basic cranial and dental pattern characterizing the hypothetical ancestral catarrhine morphotype. The skull of the living hylobatids, for example, provides in many respects a close approximation to the primitive catarrhine pattern from which the more specialized great ape/human and cercopithecoid cranial patterns are derived. Nevertheless, several shared derived catarrhine characters of the cranium and dentition, including the increased size of the incisors relative to the molars and a corresponding increase in the anterior breadth of the palate and mandible to accommodate the expanded incisors, a slight reduction of the degree of specialization of the P_3 for a sectorial function, slight modifications in the structure and proportions of the upper and lower cheek teeth, and a decrease in the size of M3, are characteristic of the extant Hominoidea (see Table 1, Node 3 for details).

Undoubtedly, the most significant hominoid specializations occur in the postcranial skeleton. A large number of shared derived features in the forelimb, hindlimb, thorax, pelvis and vertebral column can be identified in the recent hominoids (Andrews, 1985; Andrews & Groves, 1976; Ankel, 1972; Ashton & Oxnard, 1963, 1964; Cartmill & Milton, 1977; Corruccini, 1975, 1977, 1978; Corruccini *et al.*, 1975, 1976; Delson & Andrews, 1975; Evans & Krahl, 1945; Groves, 1972; Harrison, 1982, 1987; Le Gros Clark, 1959; Lewis, 1965, 1969, 1971, 1974; Lewis *et al.*, 1970; Napier & Davis, 1959; Rose, 1975, 1983, in press; Schultz, 1930, 1936, 1938, 1960, 1961, 1969; Schultz & Straus, 1945; Straus, 1963) (See Table 1, Node 3 for details). However, this extensive list includes many individual postcranial characters which are functionally related. It is, perhaps, more appropriate, therefore, to consider the characters which define the ancestral hominoid morphotype in terms of discrete functional complexes. These can be summarized as follows: (1) an increased potential for raising the forelimb above the head, (2) an increased potential for extension of the forelimb at the elbow joint, (3) a greater potential for rotation of the forelimb, (4) a greater flexibility of the wrist, (5) a more erect posture during locomotion and feeding, (6) a greater mobility at the hip and ankle joints, and (7) differential usage of the forelimb over the hindlimb (Corruccini, 1975; Harrison, 1982, 1987). These structural/functional modifications of the postcranium in hominoid primates are almost certainly associated with changes in locomotor behavior from a generalized arboreal quadrupedal running and walking mode of progression to one including more active quadrupedal climbing, bridging and forelimb suspension.

The shared derived characters which distinguish the hominoids from the ancestral catarrhine morphotype are so pervasive and detailed in their nature that it seems most unlikely that they arose independently in the extant members of the group as a result of convergent evolution (*contra* Le Gros Clark & Thomas, 1951; Napier & Davis, 1959; Simons, 1967; Pilbeam, 1969). It seems most reasonable to assume, therefore, unless we accept a remarkable degree of homoplasy, that the postcranial specializations shared by recent hominoids have been inherited from their last common ancestor (e.g., Groves, 1972; Corruccini, 1975, 1978).

Early fossil catarrhines

Having constructed the ancestral morphotypes for the major groups of the Anthropoidea based on neontological evidence, we can now attempt to examine the possible relationships between the early fossil catarrhines and the recent taxa. The best documented paleontological evidence for the initial radiation of the catarrhines comes from the

Oligocene deposits of the Fayum in Egypt and the early and middle Miocene sites of Western Kenya. This material therefore provides the basis for the review presented here. In addition, the later Miocene pliopithecids from Eurasia appear to represent an early derivative of the basal catarrhine radiation and so are included in the present analysis.

Parapithecids

The most common primates from the early Oligocene deposits of the Fayum in Egypt are the parapithecids. They are currently assigned to five species, *Qatrania wingi, Apidium moustafai, Apidium phiomense, Parapithecus fraasi* and *Simonsius grangeri* (Osborn, 1908; Schlosser, 1910; Simons, 1962, 1974; Gingerich, 1978; Simons & Kay, 1983; Fleagle, 1986). Most species are poorly known, but *Simonsius grangeri* and *Apidium phiomense* are represented by a large sample of isolated teeth, some craniodental fragments and several postcranial specimens (Simons, 1959, 1962, 1968, 1972, 1974; Gingerich, 1973; Conroy, 1976; Fleagle & Simons, 1979, 1983; Szalay & Delson, 1979; Kay & Simons, 1983; Fleagle & Rosenberger, 1983).

Although the parapithecids demonstrate a considerable diversity of dental adaptations (Simons, 1972; Kay, 1977; Kay & Simons, 1983), they share several specializations which suggest that they represent a distinct monophyletic group (Delson & Andrews, 1975; Szalay & Delson, 1979). These specializations include an enlarged P_2, upper premolars with a well-developed paraconule, waisted lower molars, upper molars with an enlarged and discrete paraconule and metaconule, greatly reduced cingula on the cheek teeth, and a mandibular corpus which increases in depth posteriorly (Simons, 1972; Delson & Andrews, 1975; Delson, 1977; Szalay & Delson, 1979).

In addition, several species of parapithecids appear to have developed some intriguing and unique specializations of the dentition and postcranium. *Simonsius*, for example, is unusual in lacking a deciduous incisor and both permanent incisors from its lower dental formula (Kay & Simons, 1983; Simons, 1986). *Apidium* retains a more conservative dental pattern with two pairs of incisors in the lower jaw, but there appears to be some justification to reaffirm the earlier conclusion (Schlosser, 1911; Kälin, 1961) that *Parapithecus fraasi* has only a single pair of lower incisors (Kay & Simons, 1983; *contra* Simons, 1972). Furthermore, *Apidium* displays a substantial apposition and syndesmosis of the distal tibio-fibular articulation; a specialized condition seen in many strepsirhines, in living and fossil tarsioids and some of the smaller ceboids (Fleagle & Simons, 1983).

The precise nature of the phylogenetic relationship of the Parapithecidae to the recent anthropoids has been the subject of considerable debate. They have been variously considered as a group ancestral to the platyrrhines (Hoffstetter, 1972, 1974, 1977, 1980; Delson, 1975), as paracatarrhines, the sister-group of all remaining catarrhines (Delson & Andrews, 1975; Delson, 1977; Szalay & Delson, 1979), as an early representative of the Cercopithecoidea (Simons, 1970, 1972, 1974; Simons in Simons & Delson, 1978), and as a paraphyletic grouping, with *Simonsius* and *Apidium* related to cercopithecoids and hominoids respectively (Kay, 1977). However, in recent years most workers have tended to consider the parapithecids as a rather specialized group of primitive anthropoids which diverged prior to the split between the modern catarrhine superfamilies (Delson & Andrews, 1975; Delson, 1977; Szalay & Delson, 1979; Harrison, 1982; Andrews, 1985; Fleagle, 1986).

The parapithecids share several important derived features with the extant Anthropoidea that indicate a close phyletic relationship. These include:

(1) The metopic suture between the frontal bones is completely obliterated in *Apidium*, as in recent anthropoids (Simons, 1959, 1972; Szalay & Delson, 1979; Fleagle & Rosenberger, 1983).

(2) An extensive postorbital septum with a large zygomatic component is present (Simons, 1959; Fleagle & Rosenberger, 1983). Szalay & Delson (1979) have argued that the postorbital closure in *Apidium* is less complete than in the modern *Tarsius*, and is therefore more conservative than in recent anthropoids. However, it is evident from CGM 26929, a partial face of *Apidium*, that postorbital closure in this taxon was at least as extensive as in modern anthropoids (see also Fleagle & Rosenberger, 1983). The postorbital septum of *Apidium* was clearly perforated by a relatively large inferior orbital fissure, as in many New World monkeys, but the specimen is too incomplete to determine whether or not a lateral orbital fissure was also present.

(3) The petrosal of *Apidium* bears evidence of an enlarged promontory artery and lacks a stapedial artery (Gingerich, 1973; Szalay & Delson, 1979; Cartmill *et al.*, 1981; Fleagle & Rosenberger, 1983), both of which are important derived characteristics of the anthropoid morphotype.

(4) The parapithecids have a fused mandibular symphysis in the midline as in all anthropoids (but this feature is also present in some strepsirhines).

(5) The lower M3 in *Apidium* and in extant anthropoids erupts after all of the deciduous premolars have been replaced (Kay & Simons, 1983).

(6) A large dorsal epitrochlear fossa is present on the posterior aspect of the medial epicondyle of the distal humerus in *Apidium* (Conroy, 1976). This depression for the attachment of the collateral ligament of the ulna is absent in strepsirhines and in *Tarsius*, but present in platyrrhines and some cercopithecoids. Its distribution in extant primates suggests that a prominent dorsal epitrochlear fossa is a synapomorphy of the Anthropoidea, that has subsequently been lost to a variable degree in many cercopithecoids and in hominoids.

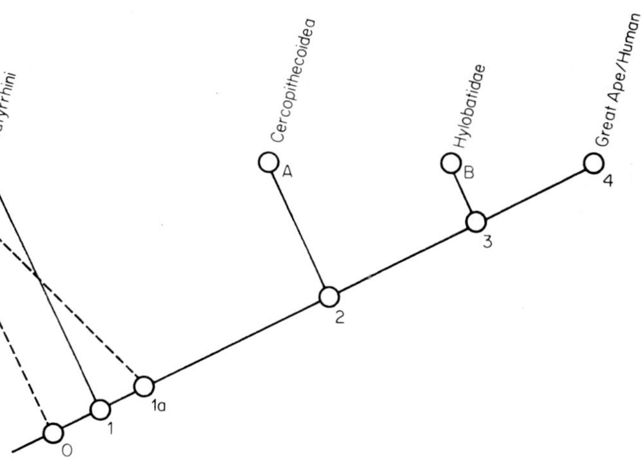

Figure 2. A cladogram illustrating the phylogenetic relationships of the Parapithecidae to recent anthropoid primates. The characters defining Nodes 0 and 1a are discussed in the text.

However, the parapithecids lack several important derived features shared by all extant anthropoid primates, which have been described here as defining the ancestral anthropoid morphotype (Node 1, Table 1). These retained primitive characters would indicate that the parapithecids should perhaps be placed on the cladogram below Node 1, as the sister group of the extant Anthropoidea (Node 0, Figure 2).

In all extant anthropoids the P_4 has two subequal cusps linked by a more or less transversely arranged crest which defines a well-developed mesial fovea and a large basin-like distal fovea. The parapithecids, however, retain a more conservative sectorial type of P_4, in which the metaconid is much smaller than the protoconid, the two cusps are separated by a longitudinal fissure, the metaconid is more markedly distally displaced relative to the protoconid, and there is little or no development of distinct mesial or distal foveae.

Furthermore, the mandible of *Apidium moustafai* (YPM 20914) has a very elevated coronoid process relative to the height of the condyle (the condyle is less than 75% of the height of the coronoid process). In most extant anthropoids the apex of the coronoid process is higher than the condyle, but the differential is much less than in *Apidium*. In addition, the development of the angular process in *Apidium*, although incomplete in YPM 20914, was apparently much greater than in any modern anthropoid, including the callitrichids, which are the only extant higher primates that still retain a distinct angular process.

The structure of P_4, the height of the coronoid process and the development of the angular process of *Apidium* more closely resemble the primitive haplorhine condition found in many tarsiiform primates than the derived pattern shared by all recent anthropoids. These retained primitive characters would seem to indicate that the parapithecids should be placed on the cladogram in Figure 1 below Node 1, as the sister group of the extant Anthropoidea. It is clear from the evidence available that *Apidium* and *Simonsius* are more conservative than any living catarrhines. They retain a number of primitive anthropoid features that are not present in extant catarrhines, such as an upper and lower P2, an annular ectotympanic and an entepicondylar foramen in the distal humerus.

The lack of several major derived catarrhine traits provides positive evidence that the parapithecids should be placed on the cladogram in Figure 1 below Node 2. This evidence contradicts the earlier suggestion, based largely on molar morphology, that the parapithecids are closely related to the cercopithecoids (Simons, 1970, 1972, 1974; Simons & Delson, 1978). The latter interpretation would imply that the cercopithecoids and hominoids have lost P2, developed a tubular ectotympanic and lost the entepicondylar foramen of the distal humerus independently. This seems highly improbable. Moreover, Delson (1975) and Szalay & Delson (1979) have argued that the supposed shared similarities between the molars of the parapithecids and cercopithecids are of questionable significance for determining their phyletic relationship.

The parapithecids share very few derived characters with the recent catarrhines. Apparent synapomorphies include the presence of wear facet x on M_2 (Kay 1977, 1980; Butler, 1986), the development of a prominent hypoconulid on M_1 and M_2 (Kay, 1980;. Butler, 1986) and the lack of a vascular canal leading from the subarcuate fossa of the sigmoid venous sinus (Cartmill *et al.*, 1981; Simons & Kay, 1983).

The presence of facet x on the distal trigonid crest is produced by occlusal contact with the mesial crest of the protocone. It is an advanced character, present in all extant catarrhines, that probably results from the mesial extension of facet 9 in molars with low trigonids (Butler, 1986). However, it is not a feature unique to catarrhines. It has been acquired independently in some platyrrhines, such as *Cebus*, and may therefore be of limited value as an indicator of phyletic affinities (Kay, 1980; Butler, 1986).

The occurrence of a well-developed hypoconulid on M_1 and M_2 is perhaps stronger evidence of a relationship with the extant catarrhines. However, the parapithecids are more conservative than the inferred ancestral catarrhine morphotype in this respect in having a relatively smaller hypoconulid, which tends to be medially or lingually placed, and in lacking a well-defined distal fovea between the hypoconulid and entoconid.

Cartmill *et al.* (1981) noted that the petrosal of *Apidium* lacks a canal connecting the subarcuate fossa to the sigmoid sinus, and this feature has been interpreted as a synapomorphy linking the parapithecids with the Catarrhini (Simons & Kay, 1983). However, the canal is only present in non-ateline ceboids among the primates, and in *Cynocephalus* among all other mammals (Cartmill *et al.*, 1981). This would suggest that the lack of a canal is a primitive primate feature, which has been retained by the parapithecids, and does not support the claim that this character is a useful synapomorphy linking the parapithecids with the catarrhines.

Fleagle & Rosenberger (1983) have further noted that the arrangement of the cranial sutures in the pterion region of *Apidium* is catarrhine-like. In extant catarrhines and in *Apidium* the alisphenoid and frontal bones are interposed between the parietal and zygomatic, while in the platyrrhines and pterion region is characterized by an extensive parieto-zygomatic junction (Ashley Montagu, 1933; Le Gros Clark, 1959). However, it is difficult to evaluate which of these alternative patterns is the primitive condition for the anthropoids. Le Gros Clark (1959), Rosenberger (1977) and Szalay & Delson (1979) consider the parieto-zygomatic junction of platyrrhines to be the specialized condition. If this is so, then the phenetic similarity of *Apidium* to extant catarrhines in the pterion region would be due to the retention of the primitive anthropoid condition. The value of this character for assessing phyletic affinities is further limited by its intraspecific variability, and by the fact that a catarrhine-like pterion region also occurs in some of the larger cebid monkeys (Ashley Montagu, 1933; Fleagle & Rosenberger, 1983).

We are left, therefore, with very slender evidence to support the catarrhine affinities of the parapithecids. The two most useful synapomorphies, the development of a relatively large hypoconulid on M_1 and M_2, and the presence of wear facet x on M_2, could well have been developed independently in the parapithecids, as they have in other mammalian lineages. On the other hand, the development of the coronoid and angular processes of the mandible, and the structure of the P_4, support the proposition that the parapithecids are not especially related to the catarrhines, and that they should more appropriately be considered as the sister group of all extant anthropoids.

With the evidence available it is difficult to place the parapithecids on the cladogram in Figure 1 with any degree of confidence. In general terms they may be considered as a specialized family of basal anthropoids. The detailed molar morphology indicates that they are the sister group of the extant Catarrhini (Delson & Andrews, 1975; Delson, 1977; Szalay & Delson, 1979; Harrison, 1982; Andrews, 1985; Fleagle, 1986), while several characters of the mandible and the lower dentition provide, in my view, slightly more persuasive evidence of a sister group relationship with all extant anthropoids. These alternative relationships are presented in the cladogram in Figure 2.

Oligopithecus

Oligopithecus savagei from the lower Fossil Wood Zone of the Fayum is known only from the holotype, a partial mandible with C-M2, and an isolated left lower M1 (Simons, 1962, 1971, 1972; Szalay & Delson, 1979; Fleagle & Kay, 1983). The dentition is characterized by an intriguing mosaic of primitive and derived anthropoid characters. The lower molars correspond almost exactly to the ancestral anthropoid pattern by retaining the following primitive features: (1) the occlusal crests and cusps are elevated and sharply defined, (2) the trigonid basin is large and slightly more elevated than the talonid, (3) the posterior margin of the trigonid lacks facet x, (4) a small, but distinct paraconid is present, at least on M_1, (5) a narrow cingulum passes around the buccal aspect of the lower molars, (6) the cristid obliqua is long and obliquely directed, (7) the hypoconulid is small and placed lingually very close to the entoconid, (8) the talonid basin is broad, shallow and expansive, (9) there is no development of a distal fovea, (10) the enamel layer is thin and unwrinkled, and (12) the M_1 is only slightly smaller than M_2 (Simons, 1972; Cachel, 1975; Kay, 1977; Szalay & Delson, 1979; Kay *et al.*, 1981; Harrison, 1982).

The lower molars of *Oligopithecus* demonstrate none of the derived characters shared by extant catarrhine primates, and in several respects (i.e., the small, lingually placed hypoconulid, the more elevated trigonid and the lack of facet x) they are more conservative than those of the parapithecids. This has led to the suggestion that *Oligopithecus* may be a primitive catarrhine of uncertain status (Delson & Andrews, 1975; Delson, 1977; Kay, 1977; Szalay & Delson, 1979; Fleagle, 1986) or even an adapid primate (Gingerich, 1977, 1980). Few recent workers have followed Simons' initial suggestion that *Oligopithecus* may represent a basal hominoid (Simons, 1962, 1972).

However, *Oligopithecus* does possess an advanced character complex, lacking in the parapithecids, that it shares with the extant catarrhines. The lower P2 has been lost in *Oligopithecus* and the P_3 has been modified into a sectorial tooth for occlusion with the upper canine. The P_3 of *Oligopithecus* is not as specialized as that in modern catarrhines, as it lacks an extension of enamel onto the mesial root, but the narrow strip of dentine exposure along its mesio-buccal face indicates that it is at least functionally equivalent. As was suggested by Delson (1975), the P_3 of *Oligopithecus* may represent an early stage in the transformation sequence leading to the more derived condition in modern catarrhines.

In addition, *Oligopithecus* has a P_4 which is more specialized than that of the parapithecids, and conforms closely to the primitive anthropoid pattern. It has two subequal cusps which are transversely aligned and the crest connecting them delimits a small mesial fovea and a broad distal fovea.

The loss of P_2 and the development of a sectorial P_3 in *Oligopithecus savagei* would appear to be a significant synapomorphy with the extant Catarrhini. However, such a relationship is contradicted by the evidence from the molar morphology of the Parapithecidae. The parapithecids share several derived features with the extant catarrhines not seen in *Oligopithecus*, which would apparently indicate that they were acquired by catarrhines prior to the loss of P2. We are therefore presented with a contradictory situation in which it is not possible to accommodate all the lines of evidence into a simple coherent evolutionary scheme. Either the enlarged hypoconulid and the development of wear facet x on the lower molars of parapithecids and recent catarrhines are parallel developments, or the P2 has been lost independently in *Oligopithecus* and extant catarrhines. Szalay & Delson (1979) have argued in favor of the latter interpretation, but as I have discussed above, there may be some sound reasons to recognize the molar adaptations of parapithecids as independent

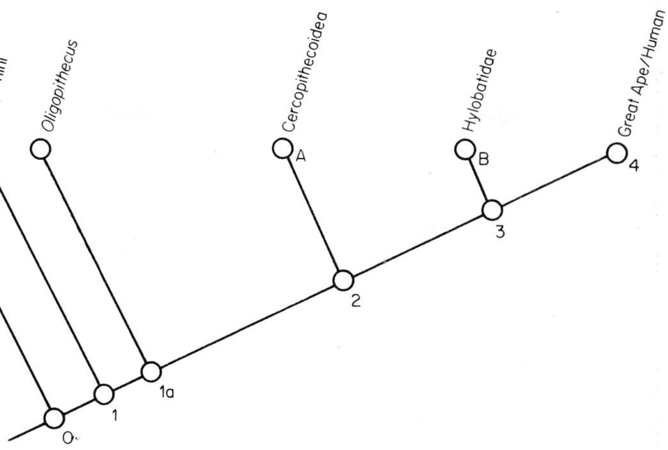

Figure 3. A cladogram illustrating the possible relationships of *Oligopithecus savagei* to the parapithecids and to the extant anthropoid primates. The characters defining the nodes are as follows: Node 0, the same as for Node 1, Table 1, except for the large angular process and the high coronoid process on the mandible, and more primitive P_4; Node 1, reduction of the angular and coronoid processes and more molariform P_4; Node 1a, loss of P_2 and development of a sectorial P_3.

acquisitions. If this is so, and the parapithecids are recognized as the sister group of the Anthropoidea, then the shared loss of P2 in *Oligopithecus* and recent catarrhines may take on greater phylogenetic significance. In my view, *Oligopithecus* provides, at present, a more reasonable sister group for the Catarrhini, based on the loss of P2, than do the parapithecids, which retain several primitive traits not seen in modern anthropoids (Figure 3). However, as *Oligopithecus* is so poorly known, a detailed, and more confident assessment of its phyletic relationships will have to await further discoveries.

Propliopithecus

The remaining anthropoid primates from the Oligocene of Egypt are generally referred to two genera, *Propliopithecus* and *Aegyptopithecus* (Kay *et al.*, 1981; Fleagle & Kay, 1983; Simons, 1985; Fleagle, 1986), although several authors have recently suggested that these two taxa may be congeneric (Szalay & Delson, 1979; Andrews, 1980, 1985). It is evident from comparisons of the dentition that *Propliopithecus* and *Aegyptopithecus* are extremely similar. They are distinguished mainly by size and by a number of minor morphological details of the cheek teeth. On this basis, there seems to be no compelling reasons for their continued inclusion in separate genera, and they are considered here to represent a single genus, *Propliopithecus*.

The largest species of *Propliopithecus, P. zeuxis*, provides the most secure basis for the assessment of the phylogenetic affinities of the genus, as its cranial, dental and postcranial morphology is much more completely known than the other three species of *Propliopithecus* (Simons, 1967, 1972, 1984, 1985; Preuschoft, 1975; Fleagle *et al.*, 1975; Conroy, 1976; Fleagle & Simons, 1978*a*, 1982; Szalay & Delson, 1979; Kay *et al.*, 1981; Fleagle, 1980, 1983*a*; Fleagle & Kay, 1983; Fleagle & Rosenberger, 1983).

Of particular importance for assessing the evolutionary relationships of the larger Oligocene anthropoids is an almost complete skull (CGM 40237) of *Propliopithecus zeuxis* (Simons, 1967, 1972; Szalay & Delson, 1979). In addition, three new faces of *P. zeuxis* have recently been recovered from the Fayum (Simons, 1984, 1985), but these have yet to be described in detail. The general structure of the cranium of *P. zeuxis* is similar in a number of respects to that of some strepsirhine primates, and contrasts with the cranial pattern inferred for the primitive anthropoid and catarrhine morphotypes. The skull of *Propliopithecus* differs from that of extant anthropoids, and resembles that of strepsirhine primates, in the following characters: (1) the rostrum is relatively long and tubular, (2) the premaxilla is relatively narrow, with a large ascending wing, (3) the zygomatic arch originates anteriorly very close to the alveolar margin of the maxilla, (4) the nasal bones, although incomplete in all specimens, were presumably relatively long and narrow, (5) the facial skeleton is relatively large compared with the size of the neurocranium [Radinsky (1979) has estimated the brain size of CGM 40237 at 27 cm^3], (6) the cranial vault is relatively low, with the frontal receding away from the orbital region at a low angle, (7) postorbital constriction is marked, (8) the neurocranium is long and narrow, with its maximum breadth situated posteriorly, and (9) the nuchal crest is robustly developed and is situated high on the skull. These individual characters relate to two major structural complexes, a large and prognathic face in association with a small neurocranium. This is certainly an unusual combination of features, not seen in any other fossil or extant anthropoid of comparable cranial size.

Nevertheless, despite the superficial similarities of the skull of *Propliopithecus* to that of some strepsirhines, a number of detailed morphological features of the cranium and mandible serve to link *Propliopithecus* with the Anthropoidea. These features include: (1) an infraorbital foramen situated close to the inferior margin of the orbit, (2) the orbit is almost completely separated from the temporal fossa by a postorbital septum, (3) the orbits are almost fully frontally directed, (4) the lacrimal bone is completely contained within the orbit, (5) the presence of an annular ectotympanic, (6) the complete fusion of the metopic suture of the frontal bone, (7) the mandibular symphysis is completely fused in the midline, (8) the mandible is deep, with a greatly reduced angular process, (9) the P_4 is bicuspid with subequal cusps and large, well-defined foveae, and (10) a large dorsal epitrochlear fossa is present on the distal humerus.

Furthermore, several features of the cranium and the detailed structure of the dentition demonstrate clearly that *Propliopithecus* shares important derived characters with the living catarrhine primates. The key features linking *Propliopithecus* with the extant catarrhines include a reduced postglenoid foramen, loss of the upper and lower P2, development of a sectorial P_3, and a general dental pattern which is described for Node 2 of Table 1. In addition, Fleagle & Rosenberger (1983) have commented on the catarrhine-like configuration of the pterion region in *Propliopithecus*, but as discussed above, the phylogenetic implications of this character are unclear.

Propliopithecus does retain, however, certain primitive anthropoid features that have been lost or modified in recent catarrhines. For example, *Propliopithecus* retains an annular ectotympanic, the primitive anthropoid condition, rather than the more specialized tubular ectotympanic characteristic of extant catarrhines. The retention of this primitive feature confirms that *Propliopithecus* should be placed on the cladogram in Figure 1 below Node 2. The phylogenetic position of *Propliopithecus*, relative to the other Fayum primates and to extant anthropoids can, therefore, be firmly established on the basis of cranial and dental characters (Figure 4).

However, there is still the problem of explaining how the distinctive facial morphology and small neurocranium of *Propliopithecus* can be accommodated within the proposed

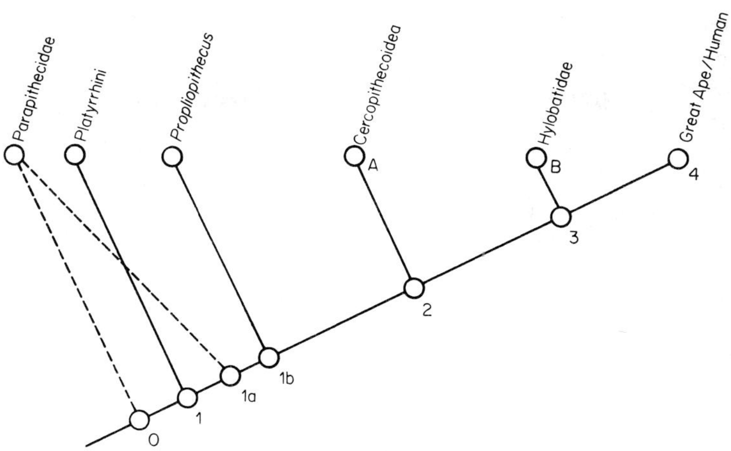

Figure 4. A cladogram depicting the phylogenetic relationships of *Propliopithecus* to the parapithecids and to the extant anthropoid primates. The characters defining Node 1b are discussed in the text.

scheme, since these appear to be incongruous with the ancestral anthropoid and catarrhine morphotypes based on extant taxa. Two hypotheses can be advanced to account for the occurrence of this apparently primitive cranial pattern in a basal catarrhine. The first hypothesis assumes that the relatively long and strepsirhine-like face and the small neurocranium of *Propliopithecus* are close to the primitive anthropoid condition. This would imply that the relatively short faces in platyrrhines and in hylobatids and many cercopithecids have been developed independently from a longer-faced ancestor. It is difficult to accept this suggestion in view of the remarkable similarity in cranial form among short-faced anthropoids, and would be even more problematic if the tarsiiform primates are accepted as the sister group of the Anthropoidea. It seems more reasonable to infer that the common ancestor of the catarrhines was a short-faced primate, with a relatively large and globular brain case (Vogel, 1966, 1968; Andrews & Delson, 1975; Szalay & Delson, 1979; Rosenberger, 1986; Strasser & Delson, 1987).

Simons (1984, 1985) has recently noted that the newly discovered faces demonstrate that *Propliopithecus* had a relatively shorter rostrum than the original discovery, and argues that *Propliopithecus* does not differ significantly from the primitive catarrhine condition in this respect (*contra* Harrison, 1982; Fleagle & Rosenberger, 1983). It is certainly true that the new faces of *Propliopithecus* appear to have relatively shorter and deeper snouts than in CGM 40237. However, when compared with most modern anthropoids of similar cranial size they still exhibit a much greater degree of facial prognathism.

The alternative proposal assumes that the cranial morphology of *Propliopithecus* is a derived anthropoid pattern in which the face has become secondarily lengthened from the hypothetical short-faced ancestral anthropoid pattern. Thus, the rather long face of *Propliopithecus* is considered to be an autapomorphy which would exclude it from direct ancestry to all later catarrhines. The first hypothesis certainly provides a tempting explanation, in view of the primitive-looking cranial morphology and the geological age of *Propliopithecus*, but the second hypothesis is much more consistent with the neontological evidence and probably provides the most acceptable interpretation.

The postcranial material of *Propliopithecus* conforms in all details to the inferred ancestral anthropoid morphotype at Node 1 (Table 1). The presence of an entepicondylar foramen in the distal humerus establishes the position of *Propliopithecus* below Node 2, and confirms its primitive catarrhine status determined from the cranial and dental evidence (Figure 4).

Propliopithecus is therefore a primitive anthropoid primate possessing a number of key structural features of the cranium and dentition which it shares with extant catarrhines. However, it lacks several important synapomorphies of the ear region and the distal humerus which unite the modern catarrhines, and it can be placed therefore below Node 2 on the cladogram in the position shown in Figure 4. The view expressed here is consistent with the phylogenetic interpretation of several recent authors (Kay *et al.*, 1981; Harrison, 1982; Andrews, 1985; Fleagle, 1986), and contradicts the earlier suggestion (Simons, 1965, 1967, 1972, 1984, 1985; Delson & Andrews, 1975; Delson, 1977; Szalay & Delson, 1979) that *Propliopithecus* is closely allied with the recent hominoids.

Pliopithecus

The pliopithecids from the middle to late Miocene of Europe are presently best assigned to seven species belonging to four genera, *Pliopithecus, Crouzelia, Anapithecus* and *Plesiopliopithecus* (Hürzeler, 1954; Zapfe, 1960, 1961; Kretzoi, 1975; Ginsburg & Mein, 1980). Recently, a new species of pliopithecid, *Laccopithecus robustus*, has been recovered from Lufeng in China (Wu & Pan, 1984, 1985), but detailed descriptions of the material are not yet available. Most of the pliopithecid species are based entirely on fragmentary cranio-dental material, but *Pliopithecus (Epipliopithecus) vindobonensis* from the Middle Miocene of Czechoslovakia is known from several partial skeletons (Zapfe, 1960, 1961; Simons & Fleagle, 1973) and these provide the basis for assessing the phyletic relationships of the Pliopithecidae.

Authors have argued in the past that *Pliopithecus* bears a close relationship to the recent gibbons (Hofmann, 1893; Hürzeler, 1954; Zapfe, 1960; Simons, 1972; Simons & Fleagle, 1973), but recent workers have presented a convincing case that these resemblances are probably due to similarity in overall size and the possession of symplesiomorphous characters (Remane, 1965; Groves, 1972, 1974; Delson & Andrews, 1975; Ciochon & Corruccini, 1977; Szalay & Delson, 1979; Ginsburg & Mein, 1980).

Cranio-dentally *Pliopithecus* is consistent with the morphological pattern defining the inferred ancestral catarrhine morphotype at Node 2 (Table 1). The partial crania of *Pliopithecus (Epipliopithecus)* conform closely to this projected ancestral pattern. The face is relatively short and broad, with a relatively wide interorbital region and large frontally directed orbits situated anteriorly above P^3–P^4. The mandibular corpus is quite low and robust, becoming shallower posteriorly, and the tooth-rows converge slightly anteriorly. The structure of the ectotympanic in *Pliopithecus* is more advanced than the annular configuration seen in the Fayum primates and in platyrrhines, and most resembles the condition in juvenile individuals of extant catarrhine primates, in which a short, partially enclosed bony tube may be developed. The morphology of the ectotympanic in *Pliopithecus* presumably represents a more primitive stage in the development of the complete bony tube of adult modern catarrhines (Zapfe, 1960; Delson & Andrews, 1975; Szalay & Delson, 1979).

The dentition of *Pliopithecus* also conforms in most details to the ancestral catarrhine morphotype at Node 2, although in several respects the dentition is uniquely specialized (e.g., the structure of the P_3, the proportions and detailed structure of the molars and the possession of a *triangle pliopithécin* between the protoconid and hypoconid on the lower molars) (Hürzeler, 1954). The dentition in *Crouzelia, Anapithecus* and *Plesiopliopithecus* appears to be more conservative than in *Pliopithecus*, retaining several features of the lower molars characteristic of the ancestral anthropoid morphotype (i.e., the crowns are much longer than broad, the occlusal crests and cusps are high and sharp, the mesial fovea is relatively large, the hypoconulid is very small, the cristid obliqua is long and obliquely directed and there is no true distal fovea) (Zapfe, 1961; Kretzoi, 1975; Ginsburg & Mein, 1980; Harrison, 1982).

The facial region, mandible and dentition display many of the derived features characteristic of the extant catarrhines, and in combination they indicate a close phyletic relationship. However, in having a partially formed tubular ectotympanic they are more primitive than any living catarrhine, but are more derived than the Fayum primates. This character indicates that *Pliopithecus* should be placed at a node above Node 1b and below Node 2 in the cladogram in Figure 4.

A comparison of the postcranial material of *Pliopithecus (Epipliopithecus)* with the inferred ancestral catarrhine morphotype provides further evidence of the status of the pliopithecids. The presence of an entepicondylar foramen in the distal humerus indicates that they are more conservative than the living catarrhines, and thus confirms their position below Node 2. Furthermore, the structure of the carpo-metacarpal joint of the thumb is a modified "hinge joint", more like the primitive anthropoid pattern seen in platyrrhines, than the more specialized "saddle" type of joint characteristic of all modern catarrhines (Napier, 1961, 1962).

The suggestion that *Pliopithecus* may represent a potential ancestral form leading exclusively to the recent hylobatids is not supported by the evidence available. The postcranial material of *Pliopithecus* conforms in almost every respect to the ancestral anthropoid and catarrhine morpotypes, and demonstrates none of the extensive range of hominoid synapomorphies. It is clear from the cranial, dental and postcranial evidence that *Pliopithecus* is a highly conservative anthropoid retaining several primitive characteristics not seen in extant catarrhines (i.e., a partially formed tubular ectotympanic, a "hinge-like" carpo-metacarpal joint of the thumb and a large entepicondylar foramen). However, the development of a short, partially enclosed tubular ectotympanic in

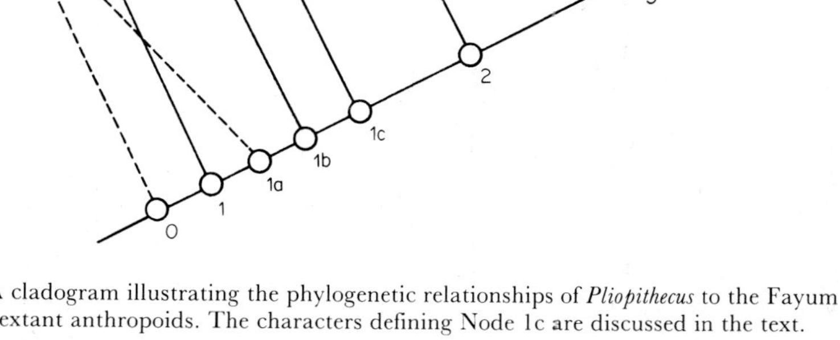

Figure 5. A cladogram illustrating the phylogenetic relationships of *Pliopithecus* to the Fayum primates and to the extant anthropoids. The characters defining Node 1c are discussed in the text.

Pliopithecus establishes a close phyletic relationship with the living catarrhines. In conclusion, *Pliopithecus* can be placed on the cladogram at Node 1c (Figure 5).

Dendropithecus

Assessment of the phylogenetic relationships of the small anthropoid primates from the early to middle Miocene of East Africa (i.e., *Limnopithecus, Dendropithecus* and *Micropithecus*) is severely limited by the lack of adequate material. Although several mandibles and fragmentary maxillae are known, and the dentition of each species is well-represented (Le Gros Clark & Leakey, 1951; Andrews & Simons, 1977; Andrews, 1978; Fleagle & Simons, 1978*b*; Harrison, 1981, 1982), important information concerning the detailed morphology of the cranium and postcranium is, in most instances, entirely lacking. For example, we have no evidence concerning the detailed structure of the basicranium, the ear region or the upper facial region of any of the species. Moreover, our knowledge of their postcranial anatomy is largely restricted to the partial skeletons of *Dendropithecus macinnesi* recovered from R3a on Rusinga Island, Kenya in 1948, and to a few isolated postcranial specimens that have provisionally been referred to other species (Le Gros Clark & Thomas, 1951; Preuschoft, 1973; Harrison, 1982; Langdon, 1986). Nevertheless, despite the inadequacies of the material, it is possible to make at least some broad inferences concerning the possible relationships of the best represented taxon, *Dendropithecus macinnesi*.

The principal concern of previous workers has been whether or not the small anthropoids from the Miocene of East Africa are phylogenetically closely related to the modern gibbons. Most early workers maintained that *Limnopithecus* (which included *Dendropithecus* at that time) was a Miocene representative of the Hylobatidae (Leakey, 1946; Le Gros Clark, 1949, 1950; Le Gros Clark & Leakey, 1950, 1951; Le Gros Clark & Thomas, 1951), and this idea was widely accepted and became firmly entrenched in the literature during the 1960s (e.g., Simons, 1963, 1965, 1972; Frisch, 1965). More recently, Andrews (1973, 1974, 1978; Andrews & Simons, 1977) has proposed that, while *Dendropithecus* is a fossil gibbon, *Limnopithecus* is more closely related to *Proconsul*, and is therefore a member of the Pongidae. By contrast, a number of other workers have argued forcefully that there is no basis to support the contention that any of the small early Miocene East African taxa are representatives of the Hylobatidae, preferring to regard them either as diminutive pongids, in the case of *Limnopithecus*, or as primitive catarrhines (Ferembach, 1958; Remane, 1965; Groves, 1972, 1974; Delson & Andrews, 1975; Delson, 1977; Harrison, 1982; Andrews, 1980, 1985; Fleagle, 1983*b*, 1986).

Cranio-dentally, *Dendropithecus* appears to conform in most details to the ancestral catarrhine morphotype. The structure of the lower face and mandible, and the morphology of the dentition, particularly the loss of P2 and the development of a sectorial P_3, indicate that *Dendropithecus* is closely related to the extant catarrhines. Unfortunately, several diagnostic features of the cranium, such as the structure of the ectotympanic, and the development of the postglenoid foramen and lateral orbital fissure, which would confirm this inferred relationship, cannot be determined in *Dendropithecus* owing to the lack of adequate material.

Of the postcranial features which define the ancestral catarrhine morphotype, only those relating to the distal humerus can be determined in *Dendropithecus*. The absence of an entepicondylar foramen is a derived feature in anthropoids which the fossil shares with all extant catarrhine primates, but which has been primitively retained by the Fayum primates and *Pliopithecus*. I have suggested previously (Harrison, 1982) that the presence of a well-defined dorsal epitrochlear fossa on the posterior aspect of the medial epicondyle of the distal humerus of *Dendropithecus* is a retained primitive feature that has been lost in all extant catarrhine primates. However, further comparisons have demonstrated that the fossa does indeed occur quite commonly in some Old World monkeys and that its absence in hominoids and most cercopithecids does not represent a useful synapomorphy defining the ancestral catarrhine morphotype.

Evidence from the distal humerus, the proximal and distal ulna, the proximal radius, the femur, the carpals and tarsals, the metacarpals and the phalanges demonstrate that *Dendropithecus*, and probably also *Limnopithecus* and *Micropithecus*, possess none of the distinctive shared derived features of the postcranium characteristic of the recent hominoids (see Table 1, Node 3 for details). There is, therefore, no convincing evidence from the postcranium, or the cranio-dental remains, to indicate a close relationship between *Dendropithecus* and the living hominoids. We can be fairly sure, therefore, that *Dendropithecus* should be placed on the cladogram in Figure 5 below Node 3 and above Node 1c. However, *Dendropithecus* is so poorly known that it is not possible to determine whether it would be placed more precisely at or below Node 2.

In conclusion, the cranio-dental evidence indicates that *Dendropithecus* and the other small catarrhine primates from the early Miocene of East Africa share a number of important derived character complexes with the extant catarrhines, placing them on the cladogram in Figure 5 above Node 1. The absence of an entepicondylar foramen on the distal humerus places *Dendropithecus* above Node 1c. Postcranially, *Dendropithecus* (and probably also *Limnopithecus* and *Micropithecus*) demonstrates a uniform absence of shared derived characters linking it specifically with either the extant hominoid primates or the Old World monkeys. This tends to restrict *Dendropithecus* to a position on the main axis of the cladogram at or below Node 2. The alternative possible positions of *Dendropithecus* on the cladogram are illustrated in Figure 6.

Proconsul

The best known genus of large catarrhine primate from the early Miocene of East Africa, *Proconsul*, has traditionally been regarded as an early representative of the African great ape

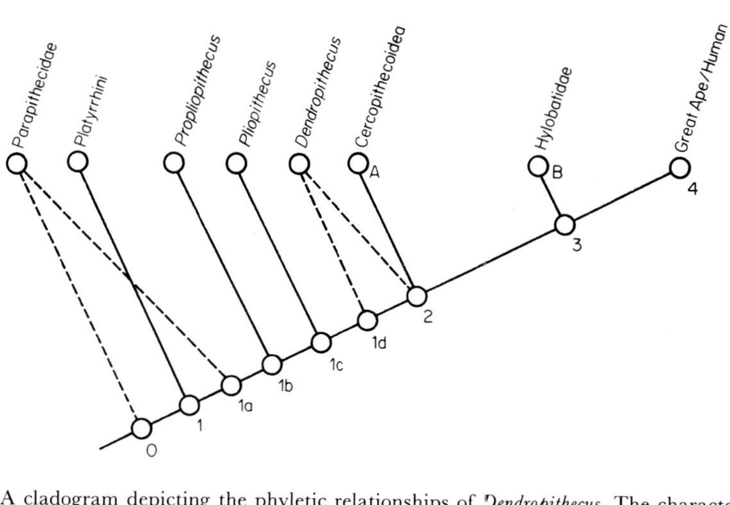

Figure 6. A cladogram depicting the phyletic relationships of *Dendropithecus*. The characters defining Node 1d and Node 2 are discussed in the text and are listed in detail in Table 1.

lineage, or more generally as a fossil "pongid" (Hopwood, 1933; Le Gros Clark, 1950; Le Gros Clark & Leakey, 1951; Napier & Davis, 1959; Pilbeam, 1969, 1972; Simons, 1972; Andrews, 1978; Szalay & Delson, 1979). However, very little evidence has been advanced to support the proposed relationship between *Proconsul* and the extant large hominoids, and several authors have recently proposed alternative schemes (Harrison, 1982; Andrews, 1985; Feldesman, 1986; Fleagle, 1986) with *Proconsul* being regarded either as a basal hominoid (e.g. Andrews, 1985) or as a more conservative anthropoid, representing the sister taxon of all extant catarrhines (e.g. Harrison, 1982).

The well-preserved cranial material of *Proconsul africanus* from Rusinga Island (Le Gros Clark & Leakey, 1951; Napier & Davis, 1959; Andrews, 1978; Walker *et al.*, 1983) corresponds in almost all cranial and dental features with the ancestral catarrhine morphotype at Node 2 (Table 1, Figure 1). These include such diagnostic catarrhine features as the possession of a well-developed tubular ectotympanic, a very reduced postglenoid foramen, specializations of the mandibular symphysis and corpus, and the loss of P2 and other derived dental features listed for Node 2 in Table 1. There appear to be no convincing derived characters of the dentition or the cranium to link *Proconsul* with either of the extant superfamilies of the Catarrhini. Many of the superficial dental and facial similarities of *Proconsul* to modern hominoids are most probably primitive retentions from the ancestral catarrhine condition. The relatively deep face and prognathic muzzle, which accounts, at least in part, for the previous suggestion that *Proconsul* may be closely related to the large hominoids, would seem to have been acquired independently as a result of convergent allometric trends in primates of similar body size (see Vogel, 1968; Shea, 1983, 1984, 1985). In my view, *Proconsul* is a relatively generalized catarrhine of large body size which shares no known derived cranial or dental traits with the extant members of the Hominoidea.

Important recent discoveries from East Africa have significantly improved our knowledge of the postcranial anatomy of *Proconsul* (Walker, 1983; Walker & Pickford, 1983; Rose, 1983; Walker *et al.*, 1986). The wealth of new material, when fully described and analyzed, will undoubtedly contribute a great deal of new information to help resolve the phyletic affinities of *Proconsul*. In the meantime, however, it is possible to present some preliminary conclusions regarding its relationships, based on the material described by Napier & Davis (1959) and on the preliminary descriptions of the new postcranial material (Walker, 1983; Walker & Pickford, 1983; Rose, 1983; Conroy & Rose, 1983; Pickford, 1985; Feldesman, 1985; Walker *et al.*, 1986).

A comparison of the postcranial morphology demonstrates that *Proconsul* conforms very closely to the hypothetical ancestral anthropoid and catarrhine morphotypes (Harrison, 1982; Andrews, 1985; Feldesman, 1986). Like the cranio-dental remains the postcranium appears to exhibit few derived anthropoid traits. The absence of an entepicondylar foramen on the distal humerus and the loss of a third trochanter on the shaft of the femur are characters which serve to link *Proconsul* with the catarrhines. In contrast, however, Napier (1961, 1962) has demonstrated that *Proconsul* is more primitive than living catarrhines in the structure of the carpo-metacarpal joint of the thumb. In non-catarrhine primates the trapezium has a cylindrical "hinge" joint, in which extension and flexion, and abduction and adduction of the thumb is possible, but axial rotation is relatively restricted (Napier, 1961, 1962). In modern catarrhines, by contrast, the carpo-metacarpal joint forms a "saddle" type of joint which permits a wide range of axial rotation of the thumb. The modification of the structure of the joint in extant catarrhines is further associated with changes in the musculature of the thumb (Day & Napier, 1963), which provides additional support that the cercopithecoids and hominoids have inherited this specialized condition from a common ancestor. The structure of the trapezium and the partial proximal epiphysis of the first metacarpal in the forelimb of *Proconsul africanus* is considered by Napier (1961, 1962) to be of a more primitive type than in recent catarrhine primates, and it most resembles the "modified hinge" type of joint seen in New World monkeys. This feature may provide good evidence to argue, as I have done previously (Harrison, 1982), that *Proconsul* is more primitive than all extant catarrhines, and should therefore be placed below Node 2 on the cladogram in Figure 1.

When the postcranial remains of *Proconsul* are compared with the ancestral hominoid morphotype (see Table 2), it is readily apparent that *Proconsul* is more primitive than any modern hominoid (Harrison, 1982; Andrews, 1985; Feldesman, 1986; Fleagle, 1986). Unless we accept a quite remarkable degree of parallelism in the postcranial skeletons of the hylobatids and the larger hominoids, it is clear that *Proconsul* can no longer be considered to have an exclusive relationship with the great apes. As discussed above, the recent hominoids share a cluster of distinctive character complexes which, apart from several individual component features of the forelimb, are entirely lacking in *Proconsul* (Table 2). Important specializations of the elbow joint to increase the potential for extension and rotation of the forelimb, the absence of an intra-articular meniscus and other specializations to increase the mobility of the wrist, and modifications of the hip, knee and ankle joints to increase flexibility in the hindlimb are all important shared derived features in the living hominoids, not present in *Proconsul*.

There are therefore justifiable reasons to place *Proconsul* on the cladogram in Figure 6 below Node 3. However, *Proconsul* conforms almost precisely with the ancestral catarrhine morphotype at Node 2, and there are few, if any, individual characters to indicate a close relationship with the recent hominoids. I have argued in the past (Harrison, 1982) that *Proconsul* shares no known derived features with the hominoids, placing it at or below Node 2. However, in a recent cladistic analysis of the fossil catarrhines, Andrews (1985), using almost the same set of character states as my earlier study, has reached a quite different conclusion. Although he agrees that *Proconsul* should be placed outside the clade represented by the modern hominoids, he has argued that *Proconsul* shares several synapomorphies with the hominoids and should, therefore, be recognized as the sister group at Node 2a in Figure 7.

Andrews (1985) has suggested that *Proconsul* shares eight hominoid synapomorphies, comprising (1) a frontal bone which is wide at bregma, but narrows anteriorly, (2) a low-crowned P_3, (3) reduced cusp heteromorphy on upper premolars, (4) the development of a maxillary jugum, (5) a scapula with an elongated vertebral border and a robust acromion process, (6) a rounded humeral head which is larger than the femoral head, (7) a humeral head which is more medially oriented, and (8) the presence of strong medial and lateral keels on the trochlea. Andrews' analysis provides the clearest and most explicitly stated case to support the hominoid status of *Proconsul*. For this reason it is worth examining in some detail the basis for this inferred relationship.

(1) *Configuration of the frontal bone:* Andrews (1985) has suggested that the recent hominoids and *Proconsul* share a derived catarrhine condition in which the frontal at bregma is relatively broad. In most platyrrhines and non-anthropoid primates the frontal bone narrows distally to form a V-shaped wedge between the parietal bones. This is presumably the primitive anthropoid condition. Among the catarrhines, this primitive

Table 2 A comparison of the postcranium of *Proconsul* with the ancestral catarrhine and hominoid morphotypes

Character state	Primitive catarrhine pattern	The condition in *Proconsul*	Primitive hominoid pattern
Humerus			
Development of lateral trochlear keel	Absent or poorly developed	Moderately well-developed	Slightly developed
Development of medial trochlear keel	Only slightly distally projecting relative to the medial epicondyle	Only slightly distally projecting	Markedly distally projecting relative to the medial epicondyle
Development of olecranon fossa	Shallow and poorly defined	Moderately deep	Deep and well-defined
Curvature of the shaft	Slight sinusoidal curvature	Slight sinusoidal curvature	Modified curvature
Torsion of the humeral head	Slight	Moderate	Moderate to marked
Ulna			
Height of the olecranon process	Relatively very high	Slightly reduced	Very reduced
Development of the trochlear notch	High and narrow	High and narrow	Low and broad
Bilateral compression of the shaft	Strong	Strong	Slight
Range of extension at the humero-ulna joint	Restricted	Restricted	Great
Development of the styloid process	Large and well-developed	Large and well-developed	Relatively small
Type of ulnar-carpal contact	Direct contact between the styloid process and carpus	Direct contact between the styloid process and carpus	Partial isolation of the styloid process from the carpus
Presence of intra-articular meniscus	Absent	Absent	Present
Extent of the radio-ulnar articulation	Restricted	Restricted	Extensive
Proportions of the distal ulna	Antero-posteriorly long and narrow	Antero-posteriorly long and narrow	Antero-posteriorly short and broad
Radius			
Shape of the radial head	Eliptical	Eliptical	Circular
Development of the styloid process	Well-developed	Well-developed	Reduced
Proportions of the radial shaft	Short and stout	Short and stout	Elongated and slender
Carpals			
Shape of the lunate	Long and narrow	Long and narrow	Relatively broad
Structure of the capitate	Proximo-distally long with narrow head	Proximo-distally long with narrow head	Proximo-distally short with large head
Type of carpo-metacarpal joint of the thumb	Saddle	Modified hinge	Saddle
Metapodials			
Shape of the distal end of the metacarpals	Heart-shaped and narrowing dorsally	Heart-shaped and narrowing dorsally	Quadrilateral and broad dorsally

Table 2—*continued*

Character state	Primitive catarrhine pattern	The condition in *Proconsul*	Primitive hominoid pattern
Femur			
Configuration of the head of the femur	Continuous with the neck posteriorly	Continuous with the neck posteriorly	Clearly distinct from the neck
Presence of a tubercle on the neck	Present	Present	Absent
Development of the trochanteric fossa	Shallow but extensive	Shallow but extensive	Deep and restricted
Symmetry of the condyles	Relatively symmetrical	Relatively symmetrical	Medial condyle larger
Long-axis of shaft	Almost vertical	Almost vertical	Tilted laterally
Proportions of the patella groove	High and narrow	High and narrow	Low and broad
Tarsals			
Proportions of the calcaneum	Long and narrow	Long and narrow	Short and broad

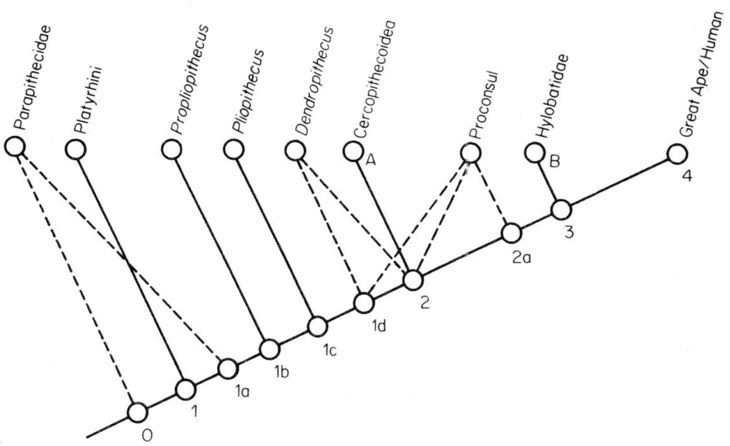

Figure 7. A cladogram illustrating the alternative phylogenetic relationships of *Proconsul*. The characters defining Node 1d and Node 2a are discussed in the text. Those defining Node 2 are summarized in Table 1.

pattern is retained in gibbons and some Old World monkeys, but most catarrhines have a fronto-parietal margin which exhibits a variable degree of posterior convexity. This convexity is least pronounced in the African apes and in humans.

The configuration of the frontal in *Proconsul* is most like that seen in some Old World monkeys and in the great apes, with a transversely broad fronto-parietal margin and a slight convexity. However, the retention of the more primitive pattern in gibbons suggests that a wide frontal bone at bregma is not a hominoid synapomorphy. If this is so, then the similarity between *Proconsul* and the great apes in this feature is unlikely to represent a shared derived trait, in view of the numerous postcranial specializations shared by all extant hominoids to the exclusion of *Proconsul*. Furthermore, the occurrence of a similar frontal pattern in many cercopithecoids reduces the utility of this feature as a useful synapomorphy linking *Proconsul* uniquely with the hominoids.

A possible explanation to account for these convergent tendencies in different anthropoid lineages is that a relationship exists between the configuration of the frontal bone and body size. Among primates the degree of penetration of the frontal between the two parietals in the sagittal plane appears to be inversely related to body size. If this is the case, then the broad frontal at bregma in *Proconsul* may simply be an expression of similar body size with the great apes rather than of a close phyletic relationship.

(2) *Relative height of P_3:* Andrews (1985) has argued that the primitive catarrhine morphotype includes a P_3 which is high-crowned relative to the length of the tooth, while *Proconsul* and the extant hominoids share a derived condition in which P_3 is relatively low-crowned. Comparisons of the P_3 in modern and fossil catarrhines do not support this view for two reasons. Firstly, *Proconsul* and the extant hominoids have a relatively higher crowned P_3 than cercopithecoids, and secondly, *Proconsul* is not significantly different in the proportions and morphology of its P_3 from that of other early catarrhine primates (Harrison, 1982). In fact, the relatively short and high-crowned P_3 in *Proconsul*, and in most non-cercopithecoid catarrhines is probably the primitive condition for the Catarrhini. It is the cercopithecoids which appear to have a derived pattern, with a relatively elongated and low P_3 and a pronounced extension of enamel onto the mesial root of the tooth. This specialization of the P_3 creates a very long honing face, particularly in males, for occlusion with the upper canine. Hominoids, by contrast, particularly the great apes, have relatively much shorter honing faces on the mesial aspect of the crown. The P_3 in *Proconsul* clearly does not resemble the extremely specialized condition in the cercopithecoids. In fact, it demonstrates an intermediate morphological pattern between cercopithecoids and hominoids, most closely resembling other fossil catarrhines from the Oligocene and Miocene. The P_3 of *Proconsul*, with a moderate degree of extension of enamel onto the mesial root of the tooth, probably represents the primitive catarrhine condition, rather than a specialized hominoid pattern.

(3) *Reduced cusp heteromorphy on the upper premolars:* Andrews (1985) claims that the primitive catarrhine morphotype is characterized by upper premolars in which there is a marked discrepancy in height between the two main cusps, while *Proconsul* and the extant hominoids have a derived condition in which the cusps are more similar in height. However, *Proconsul* displays a considerable difference in height between the cusps on P^3 and P^4, which is more pronounced than in any extant catarrhines, with the exception of the hylobatids and some cercopithecoids. It certainly does not resemble the derived pattern seen in the extant great apes. A similar disparity between the heights of the two cusps also occurs in most other early catarrhines. It would seem more probable, therefore, that the marked cusp heteromorphy seen in the hylobatids, *Proconsul* and in other fossil catarrhines represents the primitive catarrhine condition.

(4) *The development of the maxillary jugum:* Modern great apes have a pronounced maxillary jugum to accommodate their robustly constructed upper canine roots. In cercopithecids and hylobatids, by contrast, which have lightly constructed and strongly bilaterally compressed upper canine roots, the facial portion of the maxilla has a smooth contour, with no pronounced external swelling. Andrews (1985) presumably considers the condition in cercopithecids and gibbons to represent the primitive catarrhine condition. However, the Old World monkeys and the hylobatids appear to have developed, presumably independently, specialized high-crowned, slender and bilaterally compressed canines. In

New World monkeys and in most early non-cercopithecoid catarrhines, which apparently retain the primitive anthropoid and catarrhine conditions respectively, the canine root is short and stout, as in the larger hominoids, and a distinct jugum is developed in the alveolar region of the maxilla. However, because platyrrhines and most early catarrhines are relatively small in size and have abbreviated rostra, the maxillary jugum is a less conspicuous feature than it is in the larger and more prognathic great apes. *Proconsul* therefore retains a primitively large maxillary jugum, as in most non-cercopithecoid and non-hylobatid anthropoids, but it has become more exaggerated in appearance, as in the great apes, because of increased facial lengthening. As was discussed earlier, the length of the face, of which the development of the maxillary jugum is apparently a related manifestation, is probably due to an overall increase in body size in both *Proconsul* and the extant great apes, and does not necessarily provide evidence of a close phyletic relationship.

(5) *Scapula with an elongated vertebral border and a robust acromion process:* A number of authors have shown that the extant hominoids have developed a suite of important specializations of the scapula that are functionally associated with a greater range of mobility at the shoulder joint and an increased potential for raising the extended arms above the head (Le Gros Clark, 1959; Schultz, 1961; Ashton & Oxnard, 1963, 1964; Oxnard, 1967, 1969; Groves, 1972; Corruccini, 1975; Cartmill & Milton, 1977). Andrews (1985) has suggested that two of the individual derived traits of hominoids, an elongated vertebral border and a robust acromion process, are also present in *Proconsul*. However, the scapula associated with the partial skeleton of *Proconsul africanus* (Walker & Pickford, 1983; Rose, 1983), does not provide convincing evidence to support such an interpretation. In the first place, the specimen is rather fragmentary, which makes comparisons extremely problematic. For example, the vertebral border of the scapula is incomplete, and it is not possible to estimate accurately its length. In the second place, the two characters of the scapula identified by Andrews (1985) as hominoid synapomorphies need to be viewed with caution when considered in a wider comparative framework. As noted by Andrews, the acromion process appears to be relatively long and stout in *Proconsul*, projecting well beyond the level of the articular surface of the glenoid fossa. Among extant catarrhines, the acromion process is very well-developed in hominoids, as in *Proconsul*, but tends to be much more abbreviated in Old World monkeys. However, platyrrhines have scapulae with relatively elongated vertebral borders and robustly projecting acromion processes. In comparison with catarrhines, they most resemble the hominoids, although in each of these characters, the condition in platyrrhines is less pronounced. It seems reasonable to assume that the pattern seen in New World monkeys represents a close approximation to the primitive anthropoid and catarrhine conditions, and that the cercopithecoids and hominoids have developed derived, but contrasting, morphologies. The hominoids have further lengthened the vertebral border of the scapula and have exaggerated the development of the acromion process, while both have undergone a reduction in cercopithecoids (see Ashton & Oxnard, 1964 and Jolly, 1967 for a functional explanation). The relatively large acromion process, and possibly also the long vertebral border of the scapula of *Proconsul*, probably represent primitive anthropoid retentions rather than hominoid synapomorphies.

(6) *The shape and size of the humeral head:* Andrews (1985) has argued that *Proconsul* shares with the extant hominoids a rounded humeral head which is larger than the femoral head. There are a number of problems with this particular feature. Firstly, the associated skeleton of *Proconsul africanus* is lacking the femoral and humeral head from both sides

(Walker & Pickford, 1983), and so no indication of their relative sizes can be deduced. Comparisons based on isolated bones do not enable an adequate estimation to be made of the relative size of the humeral head. Secondly, it would seem that the humeral head is relatively larger than the femoral head in all catarrhines, although it is fair to say that hominoids tend to have relatively larger humeral heads than do cercopithecids (Harrison, 1987). Finally, it is not possible to determine the shape of the humeral head in *Proconsul*, as no humeri preserving the head have yet been assigned to this taxon.

(7) *Orientation of the humeral head:* Andrews (1985) has noted that *Proconsul*, like the recent hominoids, has a humerus with a medially oriented head. Although the humeral head is lacking in the *Proconsul* skeleton, it is possible to determine, from the proximal portion of the shaft and from the epiphyseal plate, that *Proconsul* displayed a moderate degree of medial torsion (Rose, 1983; Walker & Pickford, 1983). This is certainly a feature in which *Proconsul* resembles the extant hominoids, and it may represent an important shared derived character. However, as noted by Rose (1983), as an isolated character, its functional and phylogenetic significance is equivocal.

(8) *The development of the trochlear keel*: A prominent median keel on the trochlea of the distal humerus is a feature that *Proconsul* shares with the extant hominoids (Napier & Davis, 1959; Harrison, 1982; Andrews, 1985). A low rounded keel is also commonly found in colobines, but it is usually poorly developed or absent in cercopithecines and in non-catarrhine primates. However, as I have discussed at some length elsewhere (Harrison, 1987), the development of a keel in non-cercopithecine primates appears to be largely determined by body size. In arboreal primates, the median keel of the trochlea becomes more pronounced as body size increases. This allometric tendency can be demonstrated in several anthropoid lineages, and appears to have been developed independently as a result of increasing body size. This implies that the similarity in the development of the trochlear keeling in *Proconsul* and in extant great apes may have been convergent acquisitions in large-bodied primates, rather than a reflection of a close phyletic relationship.

We are therefore left with slender evidence to support the conclusion that *Proconsul* is the sister group of the extant Hominoidea. Of the features advanced by Andrews (1985) as shared derived characters linking *Proconsul* with the hominoids (Node 2a, Figure 7), only the orientation of the head of the humerus remains as a potentially significant synapomorphy. In my view, there is no convincing morphological evidence to firmly place *Proconsul* above Node 2 as the sister group of the Hominoidea. *Proconsul* conforms in almost every detail with the primitive catarrhine morphotype, and may be most appropriately placed, at least as a conservative measure, at Node 2 in Figure 7. However, there is contrary evidence from the carpo-metacarpal joint of the thumb that indicates that *Proconsul* is more primitive than all extant catarrhines, and that it should be placed at Node 1d in Figure 7.

Victoriapithecus

The oldest fossil anthropoids with undoubted affinities to the living cercopithecoid primates are from the early Miocene of Africa. These consist of an upper molar, an isolated upper canine and parts of an associated forelimb from Napak, Uganda (Pilbeam & Walker, 1968; Pickford *et al.*, 1986; Pickford & Senut, in press), three fragmentary and poorly preserved mandibles of *Prohylobates tandyi* from Wadi Moghara, Egypt (Fourtau,

1918; Simons, 1969, 1972), a single mandibular fragment of *Prohylobates simonsi* from Gebel Zelten (Delson, 1979*a*; Szalay & Delson, 1979), and several jaw fragments and some isolated teeth from Buluk, Kenya assigned to *Prohylobates* sp. (Walker & Leakey, 1984; Leakey, 1985). The largest collection of early cercopithecoids, however, is from the middle Miocene locality of Maboko Island in Kenya. The material, including mostly isolated teeth, several jaw fragments and some isolated postcranial remains has been referred to two species of *Victoriapithecus* (Von Koenigswald, 1969; Simons, 1972; Delson, 1975, 1979*b*; Szalay & Delson, 1979). Recent collections at Maboko Island have yielded extensive new cranio-dental and postcranial remains of *Victoriapithecus* (Pickford, 1986; Benefit & Pickford, 1986; Benefit, 1987), but detailed descriptions have yet to be published (Benefit, in prep.; Harrison, in prep.).

Victoriapithecus clearly shares a number of significant synapomorphies with the extant catarrhines (see Table 1, Node 2). These include: (1) a mandibular corpus which shallows posteriorly, (2) the loss of the upper and lower P2, (3) the development of a sectorial P_3 for upper canine honing, (4) a well-developed hypocone on the upper molars, (5) moderately high and narrow incisors, (6) loss of the paraconid on the lower molars, (7) M_3 much larger in size than M_2, and (8) the loss of the entepicondylar foramen in the distal humerus. Morphologically, *Victoriapithecus* demonstrates a combination of features retained from the primitive catarrhine morphotype at Node 2, and an extensive suite of specializations which it shares with the modern Old World monkeys at Node A. There are certainly no apparent primitive anthropoid characters retained by *Victoriapithecus* that would indicate a placement on the cladogram in Figure 8 below Node 2.

Victoriapithecus shares several apomorphic characters with the extant cercopithecids that provide convincing evidence of a close phyletic relationship. These synapomorphies include: (1) a mandibular symphysis which is buttressed by a robust inferior transverse torus, (2) a very restricted maxillary sinus, (3) an increased number of infraorbital foramina, (4) relatively narrow and high-crowned incisors that are oriented slightly procumbently in the jaw, (5) high-crowned, slender and strongly sexually dimorphic canines, (6) an upper canine in males which has a distinct mesial groove extending onto the root, (7) the upper premolars are relatively long, quite narrow and D-shaped, (8) upper and lower molars with a specialized bilophodont pattern (including long and narrow crowns with prominent basal flaring, buccolingual waisting, elevated cusps and crests arranged to form a lophodont pattern, mirror-imaging of occluding teeth and a reduced cingulum), (9) a greatly reduced hypoconulid on M_1 and M_2, (10) a modified crista obliqua on the upper molars, (11) a long, narrow and highly sectorial P_3, with a prominent extension of enamel onto the mesial root of the tooth, and exhibiting a high degree of sexual dimorphism, (12) an abbreviated and posteriorly directed medial epicondyle of the distal humerus, (13) a deep and well-defined olecranon fossa of the humerus, and (14) short and relatively stout phalanges.

However, *Victoriapithecus* is more primitive than any living cercopithecid in a number of features of its dentition (see Leakey, 1985 and Benefit, 1987). The upper molars and dP^4, for example, are relatively broad and are incompletely bilophodont, retaining in many cases a distinct crista obliqua. Similarly, the lower molars retain a small, but distinct hypoconulid, at least on dP_4 and M_1 and occasionally also on M_2. In addition, the metaconid of P_3 is much better developed than in modern Old World monkeys, and the lower P4 is relatively shorter and broader, with a more obliquely oriented long axis. In these respects, *Victoriapithecus* retains a more conservative morphology than any living cercopithecid and can therefore be placed on the cladogram at Node 2b (Figure 8) as the sister taxon of the modern Cercopithecidae.

Conclusions

The cladogram in Figure 8 provides a summary of the inferred phylogenetic relationships among the early catarrhine primates and the extant anthropoids. It represents a conservative statement of their phyletic affinities, in which different hypotheses have been accommodated where contrary evidence is available to support alternative interpretations. Of the different phylogenetic schemes illustrated in Figure 8, I prefer the set of inferred relationships presented and summarized in Figure 9, as it is most consistent with the fossil evidence presently available (Table 3).

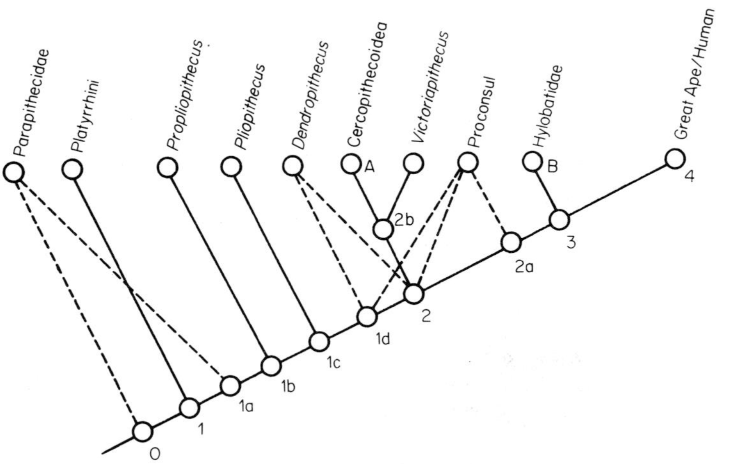

Figure 8. A cladogram illustrating the evolutionary relationships of *Victoriapithecus* to other fossil and extant anthropoids. The characters defining Node 2b are discussed in the text.

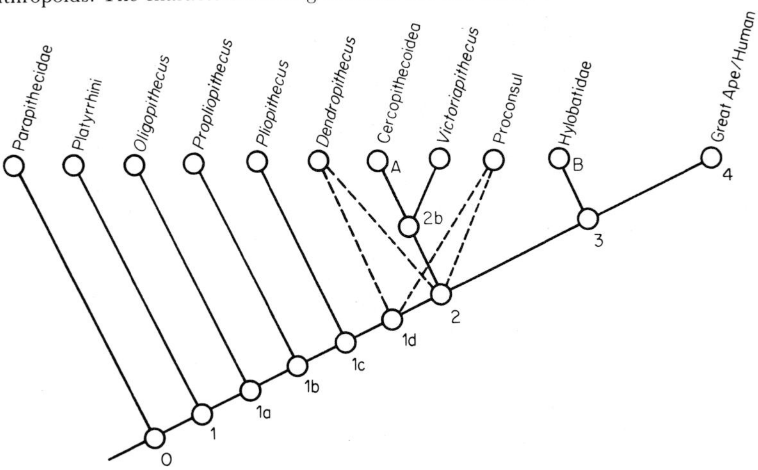

Figure 9. A cladogram summarizing the phylogenetic relationships of the early catarrhine primates most consistent with the fossil evidence presently available. The characters defining the key nodes are listed in Table 3.

Table 3 **A list of key characters defining the relevant nodes in Figure 9**

Node 0	
As for Node 1 in Table 1, except:	
31.	The angular process of the mandible is strongly developed.
32.	The coronoid process is very much higher than the condylar process.
44.	Lower P4 long and narrow with a large protoconid, a small distally displaced metaconid and no development of a mesial or distal fovea.
Node 1	
31.	The angular process of the mandible is reduced.
32.	The coronoid process is reduced in height relative to the condyle.
44.	Lower P4 bicuspid, ovoid in outline and broader than long, with subequal cusps and large foveae.
Node 1a	
33.	Dental formula 2.1.2.3., with loss of the upper and lower P2.
43.	Lower P3 specialized for occlusal contact with the upper canine.
Node 1b	
5.	Complete closure of the interpremaxillary suture early in life.
18.	Postglenoid foramen very reduced.
29.	Mandible shallows posteriorly.
30.	Mandibular tooth-rows converge only slightly anteriorly.
31.	No development of an angular process on the mandible.
38–51.	Dentition as described in Table 1, Node 2.
Node 1c	
20.	Partially formed tubular ectotympanic.
Node 1d	
20.	Fully formed tubular ectotympanic.
67.	Loss of entepicondylar foramen of distal humerus.
Node 2	
77.	Carpo-metacarpal joint of the thumb of "saddle" type.
Node 2b	
6.	Nasal aperture high and narrow with V-shaped inferior margin.
15.	Maxillary sinus greatly restricted.
28.	Mandibular symphysis with dominant inferior transverse torus.
37.	Upper canines high-crowned, strongly bilaterally compressed, with a single mesial groove on the root in males.
38.	Upper premolars relatively long and D-shaped.
39.	Upper and lower molars high-crowned long and narrow and modified to form a bilophodont pattern. Crista obliqua commonly retained.
40.	Incisors relatively narrow and very high-crowned. Slightly procumbent.
41.	Lower canines high-crowned and slender.
43.	Lower P3 long and narrow with a marked extension of enamel onto the mesial root of the tooth. Highly sectorial.
45.	Lower M1 and M2 with reduced hypoconulid.
67.	Humerus with the following morphological features: Medial epicondyle slightly posteriorly directed and abbreviated. Olecranon fossa relatively deep and well-defined.
82.	Phalanges relatively short and quite stout.
Node A	
39.	No distinct crista obliqua on the upper molars.
44.	Lower P4 longer than broad.
45.	Loss of the hypoconulid on the lower M1 and M2.
Node 3	
As for Node 3 in Table 1.	

Table 4 **Proposed classification of the fossil anthropoids discussed in this paper**

Suborder Anthropoidea Mivart, 1864
  Infraorder *incertae sedis*
    Superfamily Parapithecoidea Schlosser, 1911
      Family Parapithecidae Schlosser, 1911
          Apidium Osborn, 1908
          Parapithecus Schlosser, 1910
          Simonsius Gingerich, 1978
          Qatrania Simons & Kay, 1983
  Infraorder Platyrrhini E. Geoffroy, 1812
    Superfamily Ceboidea Bonaparte, 1831
      Family Callitrichidae Thomas, 1903
      Family Cebidae Bonaparte, 1831
  Infraorder Catarrhini E. Geoffroy, 1812
    Superfamily Propliopithecoidea Straus, 1961
      Family Propliopithecidae Straus, 1961
          Propliopithecus Schlosser, 1910
    Superfamily Pliopithecoidea Zapfe, 1960
      Family Pliopithecidae Zapfe, 1960
          Pliopithecus Gervais, 1849
          Plesiopliopithecus Zapfe, 1960
          Crouzelia Ginsburg, 1975
          Anapithecus Kretzoi, 1975
          Laccopithecus Wu & Pan, 1984
    Superfamily Proconsuloidea Leakey, 1963
      Family Proconsulidae Leakey, 1963
          Proconsul Hopwood, 1933
          Limnopithecus Hopwood, 1933
          Rangwapithecus Andrews, 1974
          Dendropithecus Andrews & Simons, 1977
          Micropithecus Fleagle & Simons, 1978
    Superfamily Cercopithecoidea Gray, 1821
      Family Cercopithecidae Gray, 1821
        Subfamily Cercopithecinae Gray, 1821
        Subfamily Colobinae Blyth 1875
        Subfamily Victoriapithecinae Von Koenigswald, 1969
          Victoriapithecus Von Koenigswald, 1969
    Superfamily Hominoidea Gray, 1825
      Family Hominidae Gray, 1825
      Family Hylobatidae Gray, 1870
    Superfamily *incertae sedis*
          Oligopithecus Simons, 1962

The status of the parapithecids remains uncertain. They may, as many workers have suggested, be the sister taxon of all other known catarrhines, but there is good evidence to suggest that they are a rather specialized family of anthropoids that diverged from the basal stock prior to the last common ancestor of all extant anthropoids (Node 0, Figure 9). If this is the case, then the loss of P2 in *Oligopithecus* may represent an important synapomorphy linking it with the extant catarrhines, despite its apparently conservative molar morphology (Node 1a, Figure 9).

The phyletic positions of *Propliopithecus* (including *Aegyptopithecus*) and *Pliopithecus* appear to be firmly established. They apparently diverged prior to the split between cercopithecoids and hominoids and they represent the successive sister taxa to the modern catarrhines (Nodes 1b and 1c, Figure 9).

So little is known of the cranial and postcranial anatomy of *Dendropithecus* that it is difficult to place it with confidence on the cladogram in Figure 9. In general terms *Dendropithecus* conforms closely with the ancestral catarrhine morphotype, and may be regarded as a basal catarrhine of modern aspect (Nodes 1d or 2, Figure 9). *Proconsul* probably occupies a similar position on the cladogram in Figure 9 to *Dendropithecus* (i.e. close to the base of the modern catarrhines). There seems to be no convincing evidence to support previous claims that *Proconsul* shares a special relationship with the hominoids (Node 2a, Figure 8). In almost all respects *Proconsul*, like *Dendropithecus*, conforms with the ancestral catarrhine morphotype at Node 2. However, the primitive structure of the carpo-metacarpal joint may place *Proconsul* below Node 2, as the sister group of all extant catarrhines (Node 1d, Figure 9).

Victoriapithecus clearly demonstrates important synapomorphies with the extant cercopithecoids (Table 3) that confirm a close phyletic relationship. However, since *Victoriapithecus* retains several primitive catarrhine features that are absent or modified in later cercopithecoids, it is recognized here as the sister taxon of extant Old World monkeys (Node 2b, Figure 9).

A preliminary classificatory scheme, based on the inferred phyletic relationships discussed in this paper, is presented in Table 4. Seven superfamilies of anthropoids are recognized in this scheme, with four distinct superfamilies, the Parapithecoidea, the Propliopithecoidea, the Pliopithecoidea and the Proconsuloidea, established to accommodate the primitive non-cercopithecid Old World anthropoids from the Oligocene and Miocene. The phyletic integrity of the Proconsuloidea is uncertain, owing to the lack of adequate fossil material for most of the included species. However, until the precise relationships among its members are resolved adequately, the Proconsuloidea is represented here, for practical purposes, as a gradistic grouping of early catarrhines of modern aspect which lack clear affinities with either of the two extant superfamilies.

Acknowledgements

I should like to thank the following colleagues and institutions for allowing me access to material in their care: R. E. Leakey and M. G. Leakey, National Museums of Kenya, Nairobi; P. J. Andrews and P. H. Napier, British Museum (Natural History), London; R. H. Tedford, G. Musser and E. Delson, American Museum of Natural History, New York; E. L. Simons and R. F. Kay, Duke University, Durham, North Carolina; R. Eissa and B. Kashab, Cairo Geological Museum, Cairo; J. Hürzeler and B. Engesser, Naturhistorisches Museum, Basel; H. Kollman and H. Zapfe, Naturhistorisches Museum, Vienna; M. Kretzoi and L. Kordos, Magyar Állami Földtani Intézet, Budapest; J-P. Lehman and L. Ginsburg, Muséum National d'Histoire Naturelle, Paris; and W. Graf, Landesmuseum "Joanneum", Gräz. I would like to thank Fred Grine, Lawrence Martin and John Fleagle for inviting me to participate in the symposium on "Primate Phylogeny" at the VIth Willi Hennig Society meeting in 1986 and to contribute to this companion volume. I am especially indebted to P. Andrews, P. Butler, E. Delson, J. G. Fleagle, C. J. Jolly, R. F. Kay, J. Kelley, M. G. Leakey, L. Martin, T. R. Olson, M. Pickford, M. D. Rose and E. Strasser for their invaluable help, advice, encouragement and comments at various stages in this research. Deborah Swartz and Kathy Rafferty ably assisted me in many ways in the preparation of the manuscript, and I thank Eugene Harris for sharing his research on the dorsal epitrochlear fossa. The research presented here was supported by grants from the Medical Research Council, London, the Boise Fund, Oxford University and the L.S.B. Leakey Foundation.

References

Andrews, P. (1973). Miocene primates (Pongidae, Hylobatidae) of East Africa. Ph.D. thesis, University of Cambridge.

Andrews, P. (1974). New species of *Dryopithecus* from Kenya. *Nature* **249,** 188–190.

Andrews, P. (1978). A revision of the Miocene Hominoidea of East Africa. *Bull. Br. Mus. (nat. Hist.) Geol.* **30,** 85–224.

Andrews, P. (1980). Ecological adaptations of the smaller fossil apes. *Z. Morph. Anthrop.* **71,** 164–173.

Andrews, P. (1981). Species diversity and diet in monkeys and apes during the Miocene. In (D. J. Chivers & K. A. Joysey, Eds) *Recent Advances in Primatology* Vol. 3, pp. 43–56. New York: Academic Press.

Andrews, P. (1985). Family group systematics and evolution among catarrhine primates. In (E. Delson, Ed.) *Ancestors: The Hard Evidence*, pp. 14–22. New York: Alan R. Liss.

Andrews, P. & Groves, C. (1976). Gibbons and brachiation. *Gibbon Siamang* **4,** 167–218.

Andrews, P. & Simons, E. L. (1977). A new African Miocene gibbon-like genus, *Dendropithecus* (Hominoidea, Primates) with distinctive postcranial adaptations: Its significance to origin of Hylobatidae. *Fol. primat.* **28,** 161–169.

Ankel, F. (1972). Vertebral morphology of fossil and extant primates. In (R. H. Tuttle, Ed.) *The Functional and Evolutionary Biology of Primates*, pp. 223–240. Chicago: Aldine.

Archibald, J. D. (1977). Ectotympanic bone and internal carotid circulation of eutherians in reference to anthropoid origins. *J. hum. Evol.* **6,** 609–622.

Ashley Montagu, M. F. (1933). The anthropological significance of the pterion in the primates. *Am. J. phys. Anthrop.* **18,** 158–336.

Ashton, E. H. & Oxnard, C. E. (1963). The musculature of the primate shoulder. *Trans. zool. Soc. Lond.* **29,** 553–650.

Ashton, E. H. & Oxnard, C. E. (1964). Functional adaptations in the primate shoulder girdle. *Proc. zool. Soc. Lond.* **42,** 49–66.

Avis, V. (1961). The significance of the angle of the mandible: An experimental and comparative study. *Am. J. phys. Anthrop.* **19,** 55–61.

Benefit, B. R. (1987). The molar morphology, natural history, and phylogenetic position of the Middle Miocene monkey *Victoriapithecus*, and their implications for understanding the evolution of the Old World monkeys. Ph.D. thesis, New York University.

Benefit, B. R. & Pickford, M. (1986). Miocene fossil cercopithecoids from Kenya. *Am. J. phys. Anthrop.* **69,** 441–464.

Butler, P. (1986). Problems of dental evolution in the higher primates. In (B. Wood, L. Martin & P. Andrews, Eds) *Major Topics in Primate and Human Evolution*, pp. 89–106. Cambridge: Cambridge University Press.

Cachel, S. (1975). The beginnings of the Catarrhini. In (R. H. Tuttle, Ed.) *Primate Functional Morphology and Evolution*, pp. 23–36. Chicago: Aldine.

Cachel, S. (1979). A functional analysis of the primate masticatory system and the origin of the anthropoid postorbital septum. *Am. J. phys. Anthrop.* **50,** 1–18.

Cartmill, M., MacPhee, R. D. & Simons, E. L. (1981). Anatomy of the temporal bone in early anthropoids with remarks on the problem of anthropoid origins. *Am. J. phys. Anthrop.* **56,** 3–21.

Cartmill, M. & Milton, K. (1977). The lorisiform wrist joint and the evolution of "brachiating" adaptations in the Hominoidea. *Am. J. phys. Anthrop.* **47,** 249–272.

Ciochon, R. L. & Corruccini, R. S. (1977). The phenetic position of *Pliopithecus* and its phylogenetic relationship to the Hominoidea. *Syst. Zool.* **26,** 290–299.

Conroy, G. C. (1976). Primate postcranial remains from the Oligocene of Egypt. *Contr. Primat.* **8,** 1–134.

Conroy, G. C. & Rose, M. D. (1983). The evolution of the primate foot from the earliest primates to the Miocene hominoids. *Foot Ankle* **3,** 342–364.

Corruccini, R. S. (1975). Morphometric affinities in the forelimb of anthropoid primates. *Z. Morph. Anthrop.* **67,** 19–31.

Corruccini, R. S. (1977). Features of the prosimian wrist joint in relation to hominoid specializations. *Acta Anat.* **99,** 440–444.

Corruccini, R. S. (1978). Primate skeletal allometry and hominoid evolution. *Evolution* **32,** 752–758.

Corruccini, R. S. Ciochon, R. L. & McHenry, H. M. (1975). Osteometric shape relationships in the wrist joint of some anthropoids. *Fol. primat.* **24,** 250–274.

Corruccini, R. S., Ciochon, R. L. & McHenry, H. M. (1976). The postcranium of Miocene hominoids: Were dryopithecines merely "dental apes"? *Primates* **17,** 205–223.

Day, M. H. & Napier, J. (1963). The functional significance of the deep head of flexor pollicis brevis in primates. *Fol. primat.* **1,** 122–134.
Delson, E. (1975). Evolutionary history of the Cercopithecidae. In (F. S. Szalay, Ed.) *Approaches to Primate Paleobiology. Contributions to Primatology,* Vol. 5, pp. 167–217. Basel: Karger.
Delson, E. (1977). Catarrhine phylogeny and classification: Principles, methods and comments. *J. hum. Evol.* **6,** 433–459.
Delson, E. (1979*a*). *Prohylobates* (Primates) from the early Miocene of Libya: A new species and its implications for cercopithecid origins. *Geobios* **12,** 725–733.
Delson, E. (1979*b*). Paleobiology of the earliest Cercopithecidae *Abstracts, VIIth. Congress of the International Primatological Society, Bangalore, India,* pp. 6–7.
Delson, E. & Andrews, P. (1975). Evolution and interrelationships of the catarrhine primates. In (W. P. Luckett & F. S. Szalay, Eds) *Phylogeny of the Primates: A Multidisciplinary Approach,* pp. 405–446. New York: Plenum.
Evans, F. G. & Krahl, V. E. (1945). The torsion of the humerus. A phylogenetic survey from fish to man. *Am. J. Anat.* **76,** 303–337.
Feldesman, M. R. (1986). The forelimb of the newly "rediscovered" *Proconsul africanus* from Rusinga Island, Kenya: Morphometrics and implications for catarrhine evolution. In (R. Singer & J. K. Lundy, Eds) *African Origins and Variation: Papers in Honour of Hertha de Villiers.* Johannesburg: Witwatersrand University Press.
Ferembach, D. (1958). Les Limnopitheques du Kenya. *Ann. Paléont.* **44,** 149–249.
Fleagle, J. G. (1980). Locomotor behaviour of the earliest anthropoids: A review of the current evidence. *Z. Morph. Anthrop.* **71,** 149–156.
Fleagle, J. G. (1983*a*). Locomotor adaptations of Oligocene and Miocene hominoids and their phyletic implications. In (R. L. Ciochon & R. S. Corruccini, Eds) *New Interpretations of Ape and Human Ancestry,* pp. 301–324. New York: Plenum.
Fleagle, J. G. (1983*b*). Are there any fossil gibbons? In (D. J. Chivers, H. Preuschoft, N. Creel, & W. Brockelman, Eds) *The Lesser Apes: Evolutionary and Behavioral Biology,* pp. 437–441. Edinburgh: Edinburgh University Press.
Fleagle, J. G. (1986). The fossil record of early catarrhine evolution. In (B. Wood, L. Martin & P. Andrews, Eds) *Major Topics in Primate and Human Evolution.* pp. 130–149. Cambridge: Cambridge University Press.
Fleagle, J. G. & Kay, R. F. (1983). New interpretations of the phyletic position of the Oligocene hominoids. In (R. L. Ciochon & R. S. Corruccini, Eds) *New Interpretations of Ape and Human Ancestry,* pp. 181–210. New York: Plenum.
Fleagle, J. G. & Rosenberger, A. L. (1983). Cranial morphology of the earliest anthropoids. In (M. Sakka, Ed.) *Morphologie Evolutive, Morphogenese du Crane et origine de l'homme.* pp. 141–153. Paris: CNRS.
Fleagle, J. G. & Simons, E. L. (1978*a*). Humeral morphology of the earliest apes. *Nature* **276,** 705–707.
Fleagle, J. G. & Simons, E. L. (1978*b*). *Micropithecus clarki,* a small ape from the Miocene of Uganda. *Am. J. phys. Anthrop.* **49,** 427–440.
Fleagle, J. G. & Simons, E. L. (1979). Anatomy of the bony pelvis in parapithecid primates. *Fol. primat.* **31,** 176–186.
Fleagle, J. G. & Simons, E. L. (1982). Skeletal remains of *Propliopithecus chirobates* from the Egyptian Oligocene. *Fol. primat.* **39,** 161–177.
Fleagle, J. G. & Simons, E. L. (1983). The tibio-fibular articulation in *Apidium phiomense,* an Oligocene anthropoid. *Nature* **301,** 238–239.
Fleagle, J. G., Simons, E. L. & Conroy, G. C. (1975). Ape limb bones from the Oligocene of Egypt. *Science* **189,** 135–137.
Fourtau, R. (1918). Contribution à l'étude des vertébres Miocenes de l'Égypte. Survey Department, Cairo.
Frisch, J. E. (1965). Trends in the evolution of the hominoid dentition. *Bibl. Primat.* **3,** 1–130.
Gingerich, P. D. (1973). Anatomy of the temporal bone in the Oligocene anthropoid *Apidium* and the origin of the Anthropoidea. *Fol. primat.* **19,** 329–337.
Gingerich, P. D. (1977). New species of Eocene primates and the phylogeny of the European Adapidae. *Fol. primat.* **28,** 60–80.
Gingerich, P. D. (1978). The Stuttgart collection of Oligocene primates from the Fayum Province of Egypt. *Paläont. Z.* **52,** 82–92.
Gingerich, P. D. (1980). Eocene Adapidae, paleobiogeography, and the origin of South American Platyrrhini. In (R. L. Ciochon & A. B. Chiarelli, Eds) *Evolutionary Biology of the New World Monkeys and Continental Drift,* pp. 123–138. New York: Plenum.
Ginsburg, L. & Mein, P. (1980). *Crouzelia rhodanica,* nouvelle espèce de primate catarhinien, et essai sur la position systématique des Pliopithecidae. *Bull. Mus. Hist. nat., Paris,* **2,** 57–85.
Groves, C. P. (1972). Systematics and phylogeny of gibbons. *Gibbon Siamang,* **1,** 1–89.
Groves, C. P. (1974). New evidence on the evolution of apes and man. *Vest. ústřed. Úst. geol.* **49,** 53–56.
Harrison, T. (1981). New finds of small fossil apes from the Miocene locality of Koru in Kenya. *J. hum. Evol.* **10,** 129–137.
Harrison, T. (1982). Small-bodied apes from the Miocene of East Africa. Ph.D. Thesis, University of London.

Harrison, T. (1987). A reassessment of the phylogenetic relationships of *Oreopithecus bambolii* Gervais. *J. hum. Evol.* **15,** 541–583.
Hershkovitz, P. (1977). *New World Monkeys (Platyrrhini)*, Vol. 1. Chicago: University of Chicago Press.
Hill, W. C. O. (1957). *Primates: Comparative Anatomy and Taxonomy.* Vol. III, *Hapalidae.* Edinburgh: Edinburgh University Press.
Hill, W. C. O. (1966). *Primates: Comparative Anatomy and Taxonomy.* Vol. VI, *Cercopithecinae.* Edinburgh: Edinburgh University Press.
Hoffstetter, R. (1972). Relationships, origins and history of the ceboid monkeys and caviomorph rodents: A modern reinterpretation. In (T. Dobzhansky, M. K. Hecht & W. C. Steere, Eds) *Evolutionary Biology* Vol. 6, pp. 323–347. New York: Appleton-Century Crofts.
Hoffstetter, R. (1974). *Apidium* et l'origine des Simiiformes (= Anthropoidea). *C. r.hebd. Séanc. Acad. Sci., Paris,* Ser. D. **278,** 1715–1717.
Hoffstetter, R. (1977). Origine et principales dichotomies des Primates Simiiformes (= Anthropoidea). *C. r.hebd. Séanc, Acad. Sci., Paris,* Ser. D. **284,** 2095–2098.
Hoffstetter, R. (1980). Origin and deployment of New World monkeys emphasizing the Southern Continents route. In (R. L. Ciochon & A. B. Chiarelli, Eds) *Evolutionary Biology of the New World Monkeys and Continental Drift,* pp. 103–122. New York: Plenum Press.
Hofmann, A. (1893). Die fauna von Göriach. *Abh. geol. Bundesanst., Wien* **15**(6).
Hopwood, A. T. (1933). Miocene primates from Kenya. *J. Linn. Soc. Lond. Zool.* **38,** 437–464.
Hürzeler, J. (1954). Contributions à l'odontologie et à la phylogénèse du genre *Pliopithecus* Gervais. *Ann. Paléont.* **40,** 1–63.
Jolly, C. J. (1967). Evolution of baboons. In (H. Vagtborg, Ed) *The Baboon in Medical Research,* Vol. 2, pp. 323–338. Austin: University of Texas Press.
Jolly, C. J. (1972). The classification and natural history of *Theropithecus (Simopithecus)* (Andrews, 1916), baboons of the African Plio-Pleistocene. *Bull. Br. Mus. (nat. Hist.), Geol.* **22,** 1–122.
Kälin, J. (1961). Sur les primates de l'Oligocène inférieur d'Egypte. *Ann. Paléontol.* **47,** 1–48.
Kay, R. F. (1977). The evolution of molar occlusion in the Cercopithecidae and early catarrhines. *Am. J. phys. Anthrop.* **46,** 327–352.
Kay, R. F. (1980). Platyrrhine origins: A reappraisal of the dental evidence. In (R. L. Ciochon & A. B. Chiarelli, Eds) *Evolutionary Biology of the New World Monkeys and Continental Drift,* pp. 159–188. New York: Plenum.
Kay, R. F., Fleagle, J. G. & Simons, E. L. (1981). A revision of the Oligocene apes of the Fayum Province, Egypt. *Am. J. phys. Anthrop.* **55,** 293–322.
Kay, R. F. & Simons, E. L. (1983). Dental formulae and dental eruption patterns in Parapithecidae (Primates, Anthropoidea). *Am. J. phys. Anthrop.* **62,** 363–375.
Kretzoi, M. (1975). New ramapithecines and *Pliopithecus* from the lower Pliocene of Rudabanya in North-Eastern Hungary. *Nature* **257,** 578–581.
Langdon, J. H. (1986). Functional morphology of the Miocene hominoid foot. *Contr. Primat.* **22,** 1–225.
Leakey, L. S. B. (1946). Fossil finds in Kenya: Ape or primitive man? *Antiquity* **20,** 201–204.
Leakey, M. (1985). Early Miocene cercopithecids from Buluk, Northern Kenya. *Fol. Primat.* **44,** 1–14.
Le Gros Clark, W. E. (1934). *Early Fore-Runners of Man.* London: Baillière.
Le Gros Clark, W. E. (1949). Early Miocene apes from East Africa. *Adv. Sci. Lond.* **5,** 340–341.
Le Gros Clark, W. E. (1950). New palaeontological discoveries bearing on the evolution of the Hominoidea. *Q. J. geol. Soc. Lond.* **105,** 225–259.
Le Gros Clark, W. E. (1959). *The Antecedents of Man.* Edinburgh: Edinburgh University Press.
Le Gros Clark, W. E. & Leakey, L. S. B. (1950). Diagnosis of East African Miocene Hominoidea. *Q. J. geol. Soc. Lond.* **105,** 260–262.
Le Gros Clark, W. E. & Leakey, L. S. B. (1951). The Miocene Hominoidea of East Africa. *Fossil Mammals Afr.* **1,** 1–117.
Le Gros Clark, W. E. & Thomas, D. P. (1951). Associated jaws and limb bones of *Limnopithecus macinnesi. Fossil Mammals Afr.* **3,** 1–27.
Lewis, O. J. (1965). Evolutionary changes in the primate wrist and inferior radio-ulnar joint. *Anat. Rec.* **151,** 275–286.
Lewis, O. J. (1969). The hominoid wrist joint. *Am. J. phys. Anthrop.* **30,** 251–268.
Lewis, O. J. (1971). Brachiation and the early evolution of the Hominoidea. *Nature* **230,** 577–578.
Lewis, O. J. (1974). The wrist articulation of Anthropoidea. In (F. A. Jenkins, Ed.) *Primate Locomotion,* pp. 143–169. New York: Academic Press.
Lewis, O. J., Hamshere, R. & Bucknell, T. (1970). The anatomy of the wrist joint. *J. Anat.* **106,** 539–552.
Napier, J. R. (1961). Prehensility and opposability in the hands of primates. *Symp. zool. Soc. Lond.* **5,** 115–132.
Napier, J. R. (1962). The evolution of the hand. *Scient. Am.* December, 56–63.
Napier, J. R. & Davis, P. R. (1959). The forelimb skeleton and associated remains of *Proconsul africanus. Fossil Mammals Afr.* **16,** 1–69.

O'Connor, B. L. (1975). The functional morphology of the cercopithecoid wrist and inferior radioulnar joints, and their bearing on some problems in the evolution of the Hominoidea. *Am. J. phys. Anthrop.* **42,** 113–122.
O'Connor, B. L. (1976). *Dryopithecus (Proconsul) africanus:* Quadruped or non-quadruped? *J. hum. Evol.* **5,** 279–283.
Osborn, H. F. (1908). New fossil mammals from the Fayum Oligocene, Egypt. *Bull. Am. Mus. nat. Hist.* **24,** 265–272.
Oxnard, C. E. (1967). The functional morphology of the primate shoulder as revealed by comparative anatomical osteometric and discriminant function techniques. *Am. J. phys. Anthrop.* **26,** 219–240.
Oxnard, C. E. (1969). Evolution of the human shoulder: Some possible pathways. *Am. J. phys. Anthrop.* **30,** 319–331.
Pickford, M. (1982). New Higher primate fossils from the Middle Miocene deposits at Majiwa and Kaloma, Western Kenya. *Am. J. phys. Anthrop.* **58,** 1–19.
Pickford, M. (1985). L'ecologie des premiers grands singes. *Recherche* **163,** 188–198.
Pickford, M. (1986). Hominoids from the Miocene of East Africa and the phyletic position of *Kenyapithecus. Z. Morph. Anthrop.* **76,** 117–130.
Pickford, M. & Senut,B. (in press). Habitat and locomotion in Miocene cercopithecoids.
Pickford, M., Senut, B., Hadoto, D., Musisi, J. & Kariira, C. (1986). Nouvelle découvertes dans le Miocène inférieur de Napak, Ouganda Oriental. *C. r.hebd. Séanc. Acad. Sci. Paris.* **302,** Ser II, 47–52.
Pilbeam, D. R. (1969). Tertiary Pongidae of East Africa: Evolutionary relationships and taxonomy. *Yale Peabody Mus. Bull.* **31,** 1–185.
Pilbeam, D. R. (1972). *The Ascent of Man.* New York: MacMillan.
Pilbeam, D. R. Walker, A. C. (1968). Fossil monkeys from the Miocene of Napak, North-East Uganda. *Nature* **220,** 657–660.
Preuschoft, H. (1973). Body posture and locomotion in some East African Miocene Dryopithecinae. In (M. Day, Ed.) *Human Evolution,* pp. 13–46. Symposium of the Society for the Study of Human Biology, Vol. XI.
Preuschoft, H. (1975). Body posture and mode of locomotion in fossil primates—method and example: *Aegyptopithecus zeuxis.* In *Proceedings of the Fifth Symposium of the Congress of the International Primatology Society,* pp. 346–359. Tokyo: Japan Science Press.
Radinsky, L. B. (1979). *Aegyptopithecus* endocast: Oldest record of a pongid brain. *Am. J. phys. Anthrop.* **39,** 239–248.
Remane, A. (1965). Die Geschichte der Menschenaffen. In (G. Herberer, Ed.) *Menschliche Abstammungstehre,* pp. 249–309. Gottingen: Gustav Fischer Verlag.
Rose, M. D. (1975). Functional proportions of primate lumbar vertebral bodies. *J. hum. Evol.* **4,** 21–38.
Rose, M. D. (1983). Miocene hominoid postcranial morphology: Monkey-like, ape-like, neither or both? In (R. L. Ciochon & R. F. Corruccini, Eds) *New Interpretations of Ape and Human Ancestry,* pp. 421–440. New York: Plenum.
Rose, M. D. (in press). Another look at the anthropoid elbow. *J. hum. Evol.*
Rosenberger, A. L. (1977). *Xenothrix* and ceboid phylogeny. *J. hum. Evol.* **6,** 561–581.
Rosenberger, A. L. (1986). Platyrrhines, catarrhines and the anthropoid transition. In (B. Wood, L. Martin & P. Andrews, Eds) *Major Topics in Primate and Human Evolution,* pp. 66–88. Cambridge, Cambridge University Press.
Schlosser, M. (1910). Über einige fossile Säugetiere aus dem Oligocän von Ägypten. *Zool. Anz.* **34,** 500–508.
Schlosser, M. (1911). Beiträge zur kenntnis dër Oligozänen Landsäugetiere aus dem Fayum: Ägypten. *Beitr. Paläont. Oest.-Ung. Orients* **6,** 1–227.
Schultz, A. H. (1930). The skeleton of the trunk and limbs of the higher primates. *Hum. Biol.* **2,** 303–438.
Schultz, A. H. (1936). Characters common to higher primates and characters specific to man. *Q. Rev. Biol.* **11,** 259–283, 425–455.
Schultz, A. H. (1938). The relative length of the regions of the spinal column in Old World primates. *Am. J. phys. Anthrop.* **6,** 1–23.
Schultz, A. H. (1960). Einige Beobachtungen und masse am Skelett von *Oreopithecus. Z. Morph. Anthrop.* **50,** 136.
Schultz, A. H. (1961). Vertebral column and thorax. *Primatologia* **4,** 1–66.
Schultz, A. H. & Straus, W. L. (1945). The number of vertebrae in primates. *Proc. Am. phil. Soc.* **89,** 601–626.
Shea, B. T. (1983). Allometry and heterochrony in the African apes. *Am. J. phys. Anthrop.* **62,** 275–289.
Shea, B. T. (1984). An allometric perspective on the morphological and evolutionary relationships between pygmy (*Pan paniscus*) and common (*Pan troglodytes*) chimpanzees. In (R. L. Susman, Ed.) *The Pygmy Chimpanzee: Evolutionary Biology and Behavior,* pp. 89–130. New York: Plenum.
Shea, B. T. (1985). Ontogenetic allometry and scaling: A discussion based on the growth and form of the skull in African apes. In (W. L. Jungers, Ed.) *Size and Scaling in Primate Biology,* pp. 175–205. New York: Plenum.
Simons, E. L. (1959). An anthropoid frontal bone from the Fayum Oligocene of Egypt: The oldest skull fragment of a higher primate. *Am. Mus. Novit.* **1976,** 1–16.
Simons, E. L. (1962). Two new primate species from the African Oligocene. *Postilla* **64,** 1–12.
Simons, E. L. (1963). A critical reappraisal of Tertiary Primates. In (J. Buettner-Janusch, Ed) *Evolutionary and Genetic Biology of Primates,* pp. 65–129. New York: Academic Press.
Simons, E. L. (1965). New fossil apes from Egypt and the initial differentiation of Hominoidea. *Nature* **205,** 135–139.

Simons, E. L. (1967). Fossil primates and the evolution of some primate locomotor systems. *Am. J. phys. Anthrop.* **26,** 241–253.
Simons, E. L. (1968). Early Cenozoic Mammalian faunas Fayum Province, Egypt, Introduction. *Peabody Mus. nat. Hist. Bull.* **28,** 1–21.
Simons, E. L. (1969). Miocene monkey (*Prohylobates*) from northern Egypt. *Nature* **223,** 687–689.
Simons, E. L. (1970). The deployment and history of Old World monkeys (Cercopithecidae, Primates). In (J. R. Napier & P. H. Napier, Eds) *Old World Monkeys,* pp. 97–137. New York: Academic Press.
Simons, E. L. (1971). Relationships of *Amphipithecus* and *Oligopithecus. Nature* **232,** 489–491.
Simons, E. L. (1972). *Primate Evolution: An Introduction to Man's Place in Nature.* New York: MacMillan.
Simons, E. L. (1974). *Parapithecus grangeri* (Parapithecidae, Old World higher primates): New species from the Oligocene of Egypt and the initial differentiation of Cercopithecoidea. *Postilla* **166,** 1–12.
Simons, E. L. (1984). Dawn ape of the Fayum. *Nat. Hist.* **95,** 18–20.
Simons, E. L. (1985). Origins and characteristics of the first hominoids. In (E. Delson, Ed.) *Ancestors: The Hard Evidence,* pp. 37–41. New York: Alan R. Liss.
Simons, E. L. (1986). *Parapithecus grangeri* of the African Oligocene: An archaic catarrhine without lower incisors. *J. human Evol.* **15,** 205–213.
Simons, E. L. & Delson, E. (1978). Cercopithecidae and Parapithecidae. In (V. Maglio & H. Cooke, Eds) *Evolution of African Mammals,* pp. 100–119. Cambridge: Harvard University Press.
Simons, E. L. & Fleagle, J. G. (1973). The history of extinct gibbon-like primates. *Gibbon Siamang* **2,** 121–148.
Simons, E. L. & Kay, R. F. (1983). *Qatrania,* a new basal anthropoid from the Fayum, Oligocene of Egypt. *Nature* **304,** 624–626.
Straus, W. L. (1963). The classification of *Oreopithecus.* In (S. L. Washburn, Ed.) *Classification and Human Evolution,* pp. 146–177. Chicago: Aldine.
Strasser, E. & Delson, E. (1987). Cladistic analysis of cercopithecid relationships. *J. hum. Evol.* **16,** 81–99.
Szalay, F. S. (1975*a*). Phylogeny of primate higher taxa: The basicranial evidence. In (W. P. Luckett & F. S. Szalay, Eds) *Phylogeny of the Primates: A Multidisciplinary Approach,* pp. 91–125. New York: Plenum.
Szalay, F. S. (1975*b*). Haplorhine relationships and the status of the Anthropoidea. In (R. H. Tuttle, Ed.) *Primate Functional Morphology and Evolution,* pp. 3–22. The Hague: Mouton.
Szalay, F. S. & Delson, E. (1979). *Evolutionary History of the Primates.* New York: Academic Press.
Temerin, L. A. & Cant, J. G. H. (1983). The evolutionary divergence of Old World monkeys and apes. *Am. Nat.* **122,** 335–351.
Vogel, C. (1966). Morphologische studien am gesichtschädel Cattarrhiner primaten. *Bibl. Primat.* **4,** 1–226.
Vogel, C. (1968). The phylogenetical evaluation of some characters and some morphological trends in the evaluation of the skull in catarrhine primate. In (A. B. Chiarelli, Ed.) *Taxonomy and Phylogeny of Old World Primates,* pp. 21–55. Turin: Rosenberg & Sellier.
Von Koenigswald, G. H. R. (1969). Miocene Cercopithecoidea and Oreopithecoidea from the Miocene of East Africa. *Foss. Vert. Afr.* **1,** 39–52.
Walker, A. (1983). The puzzle of *Proconsul. Sciences* **23,** 22–23.
Walker, A., Falk, D., Smith, R. & Pickford, M. (1983). The skull of *Proconsul africanus:* reconstruction and cranial capacity. *Nature* **305,** 525–527.
Walker, A. & Leakey, R. E. F. (1984). New fossil primates from the lower Miocene site of Buluk, N. Kenya. *Am. J. phys. Anthrop.* **63,** 232.
Walker, A. & Pickford, M. (1983). New postcranial fossils of *Proconsul africanus* and *Proconsul nyanzae.* In (R. L. Ciochon & R. S. Corruccini, Eds). *New Interpretations of Ape and Human Ancestry,* pp. 211–238. New York: Plenum.
Walker, A., Teaford, M. F. & Leakey, R. E. (1986). New information concerning the R114 *Proconsul* site, Rusinga Island, Kenya. In (J. G. Else & P. C. Lee, Eds) *Primate Evolution,* pp. 143–149. Cambridge: Cambridge University Press.
Wu Rukang & Pan Yuerong (1984). A late Miocene gibbon-like from Lufeng, Yunnan Province. *Acta Anthrop. sin.* **3,** 193–200.
Wu Rukang & Pan Yuerong (1985). Preliminary observations on the cranium of *Laccopithecus robustus* from Lufeng, Yunnan with reference to its phylogenetic relationship. *Acta Anthrop. sin.* **4,** 7–12.
Zapfe, H. (1960). Die Primatenfunde aus der Miözanen Spaltenfüllung von Neudorf an der March (Devinzka Nova Ves), Tschechoslowakei. *Schweiz. Pal. Abh.* **78,** 4–293.
Zapfe, H. (1961). Ein Primatenfund aus der Miözanen Molasse von Oberösterreich. *Z. Morph. Anthrop.* **51,** 247–267.

Elizabeth Strasser

Department of Anthropology, Graduate School, City University of New York, New York, N.Y. 10036, U.S.A.

Eric Delson

Department of Anthropology, Lehman College and the Graduate School, CUNY; and the Department of Vertebrate Paleontology, American Museum of Natural History, New York, N.Y. 10024, U.S.A.

Received 2 February 1987 Revision received and accepted 14 July 1987.
Publication date November 1987

Keywords: Morphotypes, Cercopithecidae, Cercopithecinae, Colobinae, Allenopithecina, foot morphology, cladistic analysis.

Cladistic analysis of cercopithecid relationships

The phyletic relationships of cercopithecid higher taxa are fairly well understood. By reference to the outgroup Hominoidea, character state distributions for a variety of dental, cranial and postcranial features are reviewed and morphotypes are presented for supra-generic taxa. Cercopithecids show 13 derived conditions relative to the eucatarrhine morphotype. Cercopithecines are further derived by eight features while the colobine morphotype shows an additional nine synapomorphies, including three newly identified pedal features. Within Cercopithecinae, the tribes Papionini and Cercopithecini are each characterized by several derived states. In turn, each tribe includes a relatively conservative and a more derived clade, recognized at the subtribal level: Macacina and Papionina (including the previously recognized Theropithecina) for Papionini; and the newly defined Allenopithecina and Cercopithecina among Cercopithecini. The distinctions within Colobinae appear comparable to those defining cercopithecine subtribes, and two taxa are accepted at that rank: the African Colobina, defined by three derived states; and the possibly paraphyletic Asian Presbytina. *Piliocolobus* shares four synapomorphies with *Procolobus* and therefore is ranked as a subgenus of that taxon. The hypothesis that *Dolichopithecus* is the direct descendant of *Mesopithecus* is weakened on the basis of contrary pedal evidence.

Journal of Human Evolution (1987) **16,** 81–99

Introduction

The Old World monkeys, Cercopithecidae, are the most successful family of living primates in terms of species diversity and distribution. Their origin was probably in the late Oligocene to early Miocene, and while their diversity was apparently low during the early to middle Miocene, they experienced an explosive differentiation by the late Miocene to early Pleistocene (Szalay & Delson, 1979—*Note*: unreferenced details are discussed in this source). The family comprises 20 living genera and subgenera with up to 77 species and 263 subspecies (Napier, 1981, 1985; see Figure 1). As the result of studies by numerous investigators, the overall relationships of cercopithecid higher taxa are rather well understood. In this paper, which is primarily designed as a review of current interpretations, a few alterations to existing classifications will be made while additional data are presented which support most previous cladistic analyses. Such a review is especially timely in light of the absence of any detailed discussion of cercopithecids in the recent compendium edited by Swindler & Erwin (1986).

The types of data which have proven most useful for higher level cercopithecid systematics are derived from studies of mainly hard tissues: the facial architecture, dentition and feet. The morphocline polarities and, indeed, distribution of many intriguing characters, such as biochemical and behavioral features or the anatomy of some soft tissues (e.g., genitalia), are not fully understood. Since these are most useful at lower taxonomic levels we will summarize here those features which permit the grouping of genera into supra-generic taxa, while commenting only briefly on the problems associated with the precise phyletic relationships of (sub)genera within such taxa.

The approach we have followed was to cull from the literature characters which have been suggested as diagnostic of cercopithecid family-group taxa, complementing these

0047–2484/87/010081 + 19 $03.00/0

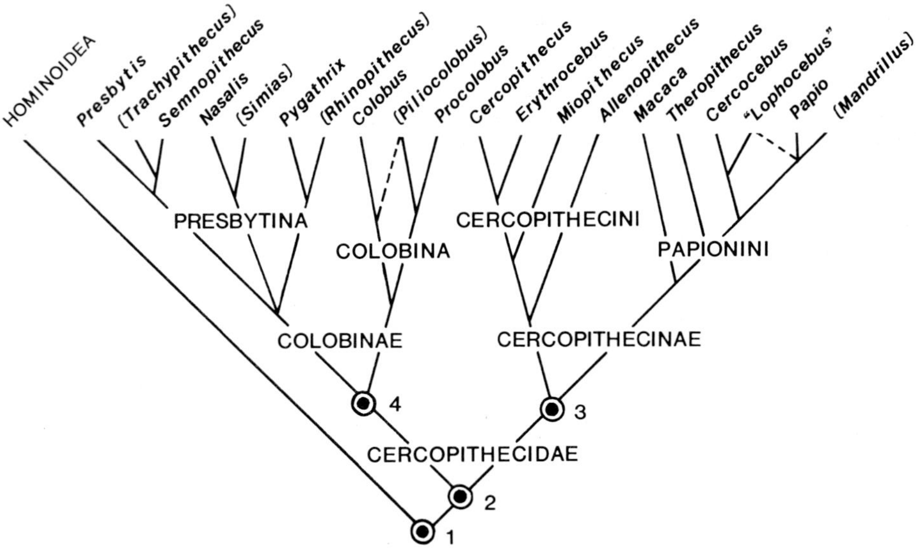

Figure 1. Cladogram of extant cercopithecid genera and subgenera. Nodes, in numerical order, identify reconstructed morphotypes for: (1) family Cercopithecidae; (2) subfamilies Colobinae and Cercopithecinae; (3) tribes Cercopithecini and Papionini; and (4) subtribes Presbytina and Colobina. Hominoidea is nearest extant sister taxon to Cercopithecidae. Subgeneric names enclosed in parentheses. "*Lophocebus*" enclosed in quotation marks to indicate controversy as to its rank.

with features currently under study. Morphocline polarity was determined for the states of each character by reference to their distribution among extant and fossil cercopithecids and in outgroups such as the Hominoidea, other catarrhines and, when necessary, platyrrhines (taxa as construed by Andrews, 1985), without reference to geochronology. Any divergence from this method will be specifically discussed. Only 37 characters were sufficiently understood to be included in this analysis, not counting those in which cercopithecids retain a conservative state by comparison to an outgroup (e.g., presence of ischial callosities, tail or narrow thorax). Of the 37 characters studied, 13 distinguish Cercopithecidae from its sister-taxon Hominoidea, and no more than nine are used to differentiate individual subtaxa (Figure 2). Since character concordancy was high, the resulting character trees were converted to the apparently most parsimonious taxon phylogeny by inspection, without utilizing numerical approaches. A separate analysis, now in progress, will evaluate alternative phylogenies using PAUP (Swofford, 1985) on the data matrix presented below.

In the following sections we review the characters at each of four major nodes which delineate Cercopithecidae and its subclades, as well as selected generic distinctions. The states for each character are identified in Table 1 and their distribution among cercopithecid (sub)genera is presented in Table 2. Each character is numbered, and in both tables and figures the postulated derived condition(s) is referenced by a primed number. The characters are not necessarily discussed individually or in numerical order, but by anatomical region, with interpretation of the transformations presented when possible.

Table 1 **Character-state definition of features analyzed**

Characters	Ancestral state (for eucatarrhines)	Derived state(s)	
inter-cusp cresting[1]	1 absent	1′ bilophodont	
dP_4–M_2 hypoconulid[1]	2 present	2′ absent	
P_3 flange[1]	3 short	3′ long	
C^1 mesial sulcus[1]	4 to cervix	4′ on root	
nasal aperture[1]	5 high and oval	5′ higher	5″ wider
post. calcaneal facet[1]	6 helical	6′ rotational	
anter. calcaneal facets[1]	7 single	7′ double	
entocuneiform groove[1]	8 strong	8′ reduced	
buccal pouches[1]	9 none	9′ present	
stomach[1]	10 simple	10′ 3-chambered	10″ 4-chambered
external pollex[1]	11 long	11′ short	11″ absent
molar flare[1]	12 none	12′ strong	12″ reduced[3]
trigonid length[1]	13 long[2]	13′ short	
molar relief[1]	14 low	14′ moderate	14″ high
I_{1-2} lingual enamel[1]	15 present	15′ absent	
nasal bone length[1]	16 short	16′ long	
lacrimal fossa[1]	17 lacrimal-maxilla	17′ lacrimal	
len. prox. ec-cu facet[1]	18 long	18′ short	18″ absent
astragalar groove[1]	19 absent	19′ present	
pedal functional axis[1]	20 digit ray 3	20′ digit rays 3–4	
M_3 hypoconulid[1]	21 present	21′ absent	
cheek tooth shape	22 subsquare	22′ elongate	
interorbital distance	23 broad	23′ narrow	
cranial vault height	24 high	24′ low	
cerebral sulci	25 "colobine-like"	25′ "cerc.-like"	25″ "hominoid"[4]
max. depth mandible	26 anterior	26′ posterior	26″ even[4]
insertion *m. flex. fib.*	27 digits 1,3,4	27′ digits 3 & 4	
face length	28 short	28′ long	
substrate preference	29 arboreal	29′ semi-terr.[3]	29″ terrestrial[4]
chromosome number	30 44?	30′ [numbers]	
anteorbital drop	31 shallow	31′ steep	
facial fossae	32 absent	32′ present	
F sexual skin cyclicity	33 present	33′ absent	
male isch. callosities	34 continuous	34′ discontinuous	
male perineal organ	35 absent	35′ present	
median mental foramen	36 variable?	36′ consistent	36″ none[4]
intermembral index	37 [numbers]		

[1] Indicates characters illustrated in Figure 3–5.
[2] See text, p. 89, for discussion of polarity alternatives.
[3] State developed at least twice in parallel (see text).
[4] State independently developed from ancestral condition, not secondarily derived.

The cercopithecid morphotype

The node distinguishing cercopithecids from hominoids can be defined by seven dental, one facial and three pedal characters. The most outstanding differences, and those that have been recognized the longest, involve the dentition, specifically the cheek teeth. The primitive condition for non-parapithecid catarrhines (eucatarrhines) is that each lower molar bears five bulbous cusps arranged around the periphery of the tooth, a condition retained in hominoids. The ancestral cercopithecid lost the hypoconulid on dP_4–M_2, elongated the cheek teeth, and realigned the cusps into a bilophodont condition (Figure 3, chars 1–2; char. 22; *Note*: characters 22–37 not illustrated). A dentition of this sort is designed for folding and slicing food, which suggested to Andrews (1981) that early cercopithecids probably included a significant proportion of leaves in their diet. In addition, the catarrhine morphotype possessed cingula on the buccal face of lower cheek teeth and the lingual face of uppers. Delson (1975) suggested that most of the cingulum was lost in cercopithecid ancestors through incorporation into the tooth sidewall, resulting in a "flared" condition wherein the tooth is widest at the cervix and narrows perceptibly toward the cusp apices (Figure 4, char. 12; see discussion below). Extant cercopithecids present cingula only as small remnants in the notches, while in the earliest fossils it is more extensive.

The ancestral eucatarrhine P_3 is reconstructed as being obliquely angled to the toothrow with its anterior surface carrying a slight flange for sharpening of the upper canine. This condition is retained or reduced in most hominoids (see Harrison, 1987), but in cercopithecids the flange is extended well below the alveolar surface of neighboring teeth providing, especially in males, an increased surface area for sharpening the dagger-like upper canines. The upper canine also has a sulcus or groove which in non-cercopithecid eucatarrhines incises its mesial surface but stops short of the cervix. In cercopithecids, this groove extends beyond the cervix of the tooth and onto the root, where it locks onto a ridge of bone in the alveolus, probably providing resistance to twisting of the tooth during intensive activity (Figure 3, char. 3–4). In terms of facial morphology, Harrison (1982) and Andrews (1985) have recognized a high, oval nasal aperture as primitive for Old World anthropoids, with cercopithecids having a derived higher and more narrow configuration (Figure 3, char. 5). This may be in accommodation to elongated canine roots, again especially in males. Additionally, Vogel (1968) argued that a consistently present median mental foramen characterized this morphotype.

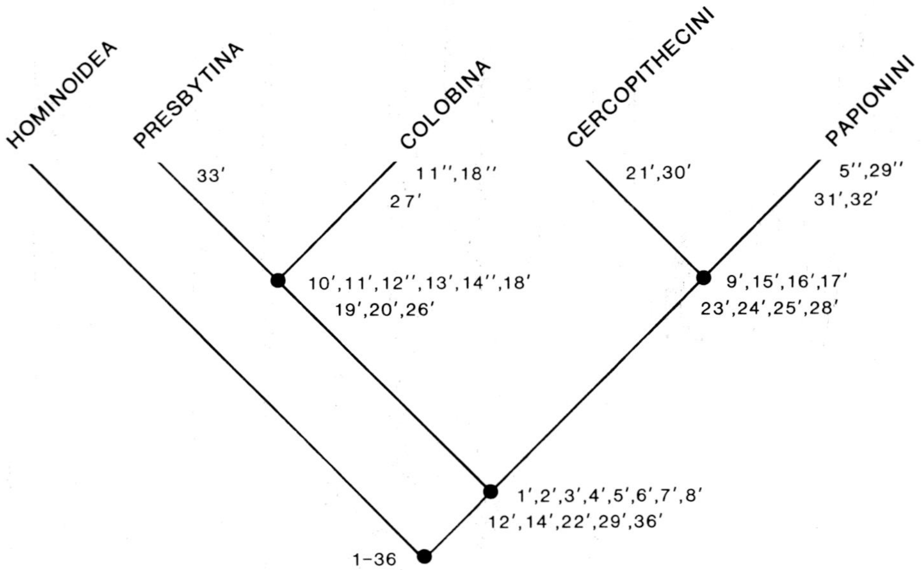

Figure 2. Character state distribution among the higher taxa of Figure 1. See Table 1 for character definitions and states, Table 2 for state definition by genus and Figures 3–5 for illustrations of selected states.

Table 2 **Data matrix for distribution of character states under discussion[1]**

Ch #	Eu Mo	Ho Mo	†Cd Mo	†Ph	†Vi	Cn Mo	Al	Mi	Er	Ce	†Ma	†Pr	†Pd	Th	Cc	Lo	Pa	Mn	†Di	†Pp	†Go	Co Mo	Co	Pc	Pi	†Li	†Ce	†Pl	†Rh	Py	Se	Tr	Pg	Na	†Me	†Do
1	1	1	1′	1′	1′	1′	1′	1′	1′	1′	1′	1′	1′	1′	1′	1′	1′	1′	1′	1′	1′	1′	1′	1′	1′	1′	1′	1′	—	1′	1′	1′	1′	1′	1′	1′
2	2	2	2′	2′	2′	2′	2′	2′	2′	2′	2′	2′	2′	2′	2′	2′	2′	2′	2′	2′	2′	2′	2′	2′	2′	2′	2′	2′	—	2′	2′	2′	2′	2′	2′	2′
3	3	3	3′	3′	3′	3′	3′	3′	3′	3′	3′	3′	3′	3′	3′	3′	3′	3′	3′	3′	3′	3′	3′	3′	3′	3′	3′	3′	—	3′	3′	3′	3′	3′	3′	3′
4	4	4	4′	—	4′	4′	4′	4′	4′	4′	4′	4′	4′	4′	4′	4′	4′	4′	4′	4′	4′	4′	4′	4′	4′	4′	4′	4′	—	4′	4′	4′	4′	4′	4′	4′
5	5	5	5′	—	—	5′	5′	5′	5′	5′	5″	5″	5″	5″	5″	5″	5″	5″	5″	5″	5″	5′	5′	5′	5′	5′	5′	5′	—	5′	5′	5′	5′	5′	5′	5′
6	6	6	6′	—	6′	6′	6′	6′	6′	6′	6′	—	—	6′	6′	6′	6′	6′	—	—	—	6′	6′	6′	6′	—	—	6′	—	6′	6′	6′	6′	6′	6′	6′
7	7	7	7′	—	7′	7′	7′	7′	7′	7′	7′	—	—	7′	7′	7′	7′	7′	—	—	—	7′	7′	7′	7′	—	—	7′	—	7′	7′	7′	7′	7′	7′	—
8	8	8	8′	—	—	8′	8′	8′	8′	8′	8′	—	—	8′	8′	8′	8′	8′	—	—	—	8′	8′	8′	8′	—	—	8′	—	8′	8′	8′	8′	8′	8′	—
9	9	9	9	—	—	9′	9′	9′	9′	9′	9′	—	—	9′	9′	9′	9′	9′	—	—	—	9	9	9	9	—	—	—	—	9	9	9	9	9	—	—
10	10	10	10	—	—	10	10	10	10	10	10	—	—	10	10	10	10	10	—	—	—	10′	10′	10″	10″	—	—	—	—	10′	10′	10′	10′	10′	—	—
11	11	11	11	—	—	11	11	11	11	11	11	—	—	11	11	11	11	11	—	—	—	11′	11″	11″	11″	—	—	—	—	11′	11′	11′	11′	11′	11′	—
12	12	12	12′	12′	12′	12′	12′	12″	12″	12″	12′	12′	12′	12′	12′	12′	12′	12′	12′	12′	12′	12″	12″	12″	12″	12″	12″	12″	—	12″	12″	12″	12″	12″	12″	12″
13	13	13	13	13?	13′	13	13	13	13	13	13	13	13	13	13	13	13	13	13	13	13	13′	13′	13′	13′	13′	13′	13′	—	13′	13′	13′	13′	13′	13′	13′
14	14	14	14′	14′	14′	14′	14′	14′	14′	14′	14′	14′	14′	14′	14′	14′	14′	14′	14′	14′	14′	14″	14″	14″	14″	14″	14″	14″	—	14″	14″	14″	14″	14″	14″	14″
15	15	15	15	—	—	15′	15′	15′	15′	15′	15′	15′	15′	15′	15′	15′	15′	15′	15′	15′	15′	15	15	15	15	—	15	15	—	15	15	15	15	15	15	15
16	16	16	16	—	—	16′	16′	16′	16′	16′	16′	—	16′	16′	16′	16′	16′	16′	16′	16′	16′	16	16	16	16	16′	16	16	—	16	16	16	16	16	16′	16
17	17	17	17	—	—	17′	17′	17′	17′	17′	17′	—	17′	17′	17′	17′	17′	17′	17′	17′	17′	17	17	17	17	17	17	17	—	17	17	17	17	17	17	17
18	18	18	18	—	—	18	18	18	18	18	18	—	—	18	18	18	18	18	—	—	—	18′	18″	18″	18″	—	—	18″	—	18″	18′	18′	18′	18′	18″	18′
19	19	19	19	—	—	19	19	19	19	19	19	—	—	19	19	19	19	19	—	—	—	19′	19′	19′	19′	—	—	19′	—	19′	19′	19′	19′	19′	19′	—
20	20	20	20	—	—	20	20	20	20	20	20	—	—	20	20	20	20	20	—	—	—	20′	20′	20′	20′	—	—	20′	—	20′	20′	20′	20′	20′	20′	—
21	21	21	21	21	21	21	21′	21′	21′	21′	21	21	21	21	21	21	21	21	21	21	21	21	21	21	21	—	21	21	—	21	21	21	21	21	21	21
22	22	22	22′	22′	22′	22′	22′	22′	22′	22′	22′	22′	22′	22′	22′	22′	22′	22′	22′	22′	22′	22′	22′	22′	22′	22′	22′	22′	—	22′	22′	22′	22′	22′	22′	22′
23	23	23	23	—	—	23′	23′	23′	23′	23′	23′	—	23′	23′	23′	23′	23′	23′	23′	23′	23′	23	23	23	23	23′	23	23	23′	23	23	23	23	23	23	23′
24	24	24	24	—	—	24′	24′	24′	24′	24′	24′	—	24′	24?	24′	24′	24′	24′	24′	24′	24′	24	24	24	24′	24′	24	24	—	24	24	24	24	24	24	24
25	25	25″	25			25′	25′	25′	25′	25′	25′	—	25′	25′	25′	25′	25′	25′	—	25′	—	25	25	25	25	25	25	25	—	25	25	25	25	25	25	25
26	26	26	26	—	26	26	26	26	26	26	26	26	—	26	26	26	26	26	26	26	26	26′	26″	26′	26′	—	26	26′	26′	26′	26′	26′	26′	26″	26″	26′
27	27	27	27	—	—	27	27	27	27	27	27	—	—	27	27	27	27	27	—	—	—	27	27′	27′	27′	—	—	—	—	27′	27	27	27	27	—	—
28	28	28	28	—	—	28′	28′	28′	28′	28′	28′	28′	28′	28′	28′	28′	28′	28′	28′	28′	28′	28	28	28	28	28′	28	28′	28′	28	28	28	28	28′	28	28′
29	29	29	29′	—	—	29′	29′	29	29″	29′	29′	29″	29″	29″	29′	29	29″	29″	—	29′	—	29′	29	29	29	—	29″	29′?	29	29	29″	29′	29	29?	29″	29″
30	44?	44?	44?	—	—	42	48	54	54	58–72	42	—	—	42	42	42	42	42	—	—	—	44	44	—	44	—	—	—	—	—	44	44	44	48	—	—
31	31	31	31	—	—	31	31	31	31	31	31′	31	31	31′	31′	31′	31′	31′	31′	31	31′	31	31	31	31	31	31	31	—	31	31	31	31	31	31	31
32	32	32	32	—	—	32	32	32	32	32	32	32	32′	32′	32′	32′	32′	32′	32	32	32′	32	32	32	32	32	32	32	—	32	32	32	32	32	32	32
33	33	33	33	—	—	33	33	33	33′	33′	33	—	—	33	33	33	33	33	—	—	—	33	33′	33	33	—	—	—	—	33′	33′	33′	33′	33′	—	—
34	34	34	34	—	—	34	34	34	34	34	34	—	—	34	34	34	34	34	—	—	—	34	34	34′	34′	—	—	—	—	34	34	34	34	34	—	—
35	35	35	35	—	—	35	35	35	35	35	35	—	—	35	35	35	35	35	—	—	—	35	35	35′	35′	—	—	—	—	35	35	35	35	35	—	—
36	36	36	36′	—	—	36′	36′	36′	36′	36′	36′	36′	36′	36′	36′	36′	36′	36′	36′	36′	36′	36′?	36″	36″	36″	—	36′	36″	36′	36″	36″	36″	36″	36″	36″	36
37	(range of values)	—	—	—	—	—	—	82–85	89–96	77–88	84–99	—	—	97	81–90	77–82	90–100	88–96	—	—	—	—	76–83	—	79–89	—	—	95	—	73–79	73–83	78–87	85–97	89–97	88	100

[1] Characters as listed in Table 1. Abbreviations as follows: († for extinct taxa) Eu Mo, eucatarrhine morphotype; Ho Mo, hominoid morphotype; Cd Mo, cercopithecid morphotype; Ph, *Prohylobates*; Vi, *Victoriapithecus*; Cn Mo, cercopithecine morphotype; Al, *Allenopithecus*; Mi, *Miopithecus*; Er, *Erythrocebus*; Ce, *Cercopithecus*; Ma, *Macaca*; Pr, *Procynocephalus*; Pd, *Paradolichopithecus*; Th, *Theropithecus*; Cc, *Cercocebus*; Lo, "*Lophocebus*"; Pa, *Papio*; Mn, *Mandrillus*; Di, *Dinopithecus*; Pp, *Parapapio*; Go, *Gorgopithecus*; Co, Mo colobine morphotype; Co, *Colobus*; Co, *Colobus*; Pc, *Procolobus*; Pi, *Piliocolobus*; Li, *Libypithecus*; Ce, *Cercopithecoides*; Pl, *Paracolobus*; Rh *Rhinocolobus*; Py, *Presbytis*; Se, *Semnopithecus*; Tr, *Trachypithecus*; Pg, *Pygathrix*; Na, *Nasalis*; Me, *Mesopithecus*; Do, *Dolichopithecus*.

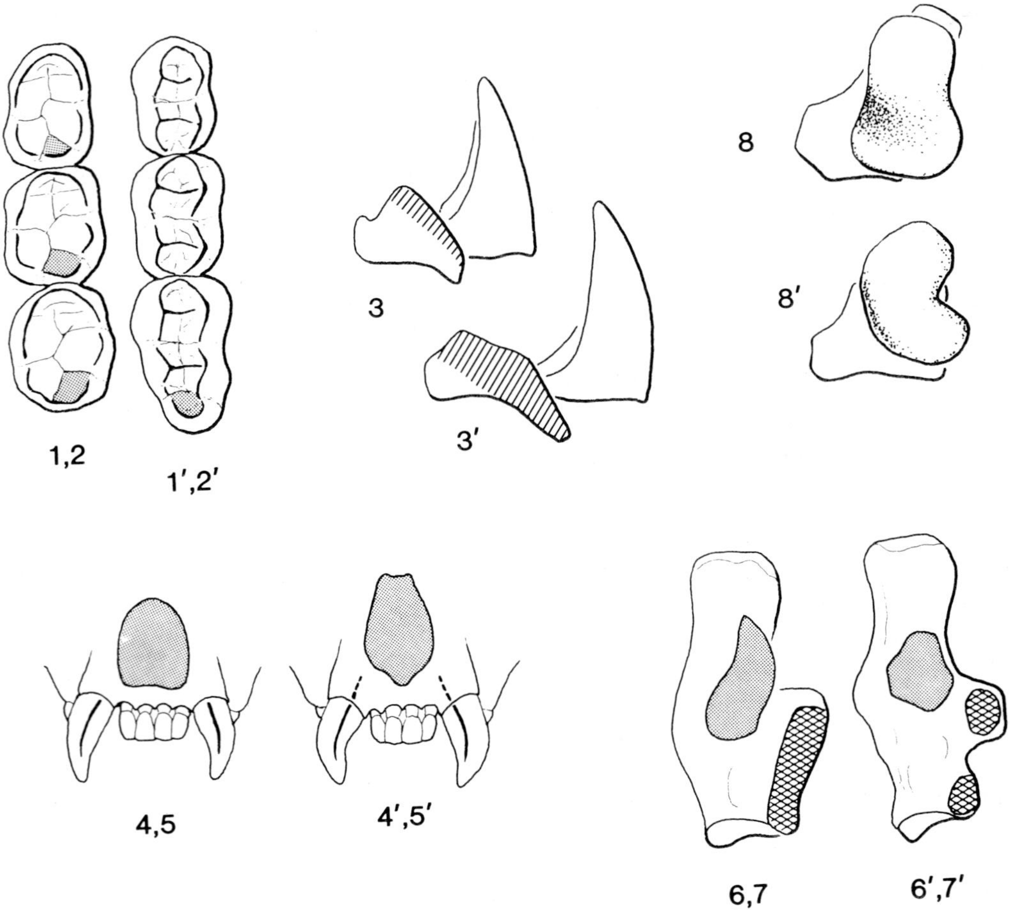

Figure 3. Selected characters at Node 1. Dorsal view RM_{1-3}, mesial to top, buccal to right: 1 peripheral bunodont cusps, 1′ bilophodont cusps; 2 hypoconulid (stippled) on all cheek teeth, 2′ hypoconulid on M_3 only. Buccal view RC_1–P_3, mesial to right: 3 short P_3 flange, moderate wear facet (parallel lines), 3′ flange extends below alveolar surface, well developed wear facet. Anterior view of lower face: 4 C^1 mesial sulcus restricted to crown, 4′ C^1 mesial sulcus extends to root (dashed lines); 5 high, oval nasal aperture (stippled), 5′ higher, narrower nasal aperture. Dorsal view right calcaneus, proximal to top, medial to right: 6 posterior astragalar facet (stippled) long and helical, 6′ posterior astragalar facet short & rotational; 7 sustentacular and anterior astragalar facets (cross-hatch) united, 7′ sustentacular (top) and anterior astragalar facets separate. Distal view left entocuneiform, plantar to bottom, proximal to left: 8 dorsal globular head, inferior helical groove, 8′ kidney-shaped, groove peripheralized.

In the foot, the morphotypic condition for eucatarrhines, and indeed for all primates, is to have a calcaneus which bears posteriorly a long, helically-curving facet and anteromedially a single facet for the astragalus, which facilitates large degrees of inversion and eversion. As recognized by several authors (Szalay, 1975; Langdon, 1986; Gebo & Simons, 1987; Rose, 1987), most hominoids retain this morphology while cercopithecids have shortened the posterior facet and have divided the anteromedial facet into separate moieties for the astragalar head and neck (Figure 3, char. 6–7). This remodelling produces a lower ankle joint which emphasizes stability at the expense of limiting inversion.

In addition, the eucatarrhine morphotype is reconstructed to have had an entocuneiform joint surface for the hallux which presented a globular dorsal head surrounded inferiorly by a helical groove. This allows the hallux to rotate on its long axis while being adducted to the other digits (Lewis, 1972; Strasser, in prep. *a, b*). The cercopithecid morphotype shows a remodelling of the entocuneiform surface into a kidney-shaped structure. The helical groove is essentially reduced and peripheralized, thereby restricting the amount of conjunct rotation that can accompany abduction–adduction (Figure 3, char. 8; Strasser, in prep. *a, b*).

This combination of characters suggests a scenario in which the ancestral cercopithecid diverged from other eucatarrhines by exploiting resources in either a more open-country environment or on the forest floor. They may have occupied a niche exploited by few hominoids, due to their pedal adaptations which could have supported rapid travel along the ground and their dentition which facilitated the inclusion of more fiber in their diet.

Colobinae *vs* Cercopithecinae

The remaining 25 characters (plus char. 12 introduced above) serve to distinguish cercopithecid subclades; in 21 cases, a primitive eucatarrhine character state has been retained from the cercopithecid morphotype. A number of these characters are directly related to dietary specializations and, as recognized for a century, distinguish cercopithecines and colobines. Cercopithecines diverge from the morphotype by developing capacious buccal pouches for the temporary storage and preparation of food (Murray, 1975; Figure 4, char. 9), about which they are highly eclectic. Colobines lack cheek pouches, but correlated with their more specialized, mostly folivorous, they have a modified stomach. Extra chambers, added to the fundus, enhance fermentation by providing more space for a highly concentrated cellulose-digesting bacterial flora (Kuhn, 1964; Bauchop & Martucci, 1968; Figure 4, char. 10).

Cercopithecines show specializations of the incisors while retaining the primitive cercopithecid molar crown configuration. The latter condition is one in which the crowns showed moderate relief, were broadly flared at the base but narrowed at the apex, and retained a relatively long trigonid (Delson, 1975; Figure 4, chars. 12–14). This group of features should not be dealt with as a "functional complex", as that would obscure the different presumptive histories of each character. The first character, crown relief, showed some increase relative to its development in most anthropoids, but not to the extreme degree seen independently developed in colobines and the *Theropithecus* clade, where deep notches incise the tooth walls.

The second character, molar flare, superficially presents a distribution among *extant* cercopithecid tribes which is counter-intuitive to the inference that its strong development is primitive for the clade, since it is found consistently only in the papionins (Figure 2, char. 12). However, *Allenopithecus*, the most primitive member of Cercopithecini (see below) also shows strong flare. Additionally, the mid-Miocene African subfamily Victoriapithecinae is not illustrated as one of the terminal taxa in Figure 2, but the species in this group also show a moderate to high degree of molar flare. Thus, strong molar flare is found in two of the three cercopithecid subfamilies. Delson's (1975) hypothesis that flare is the result of cingulum incorporation could be tested by means of embryological studies, although this has not been done to date, and Benefit (1987) has argued against it. Nevertheless, the conclusion drawn from the distribution within the family is supported by reference to an

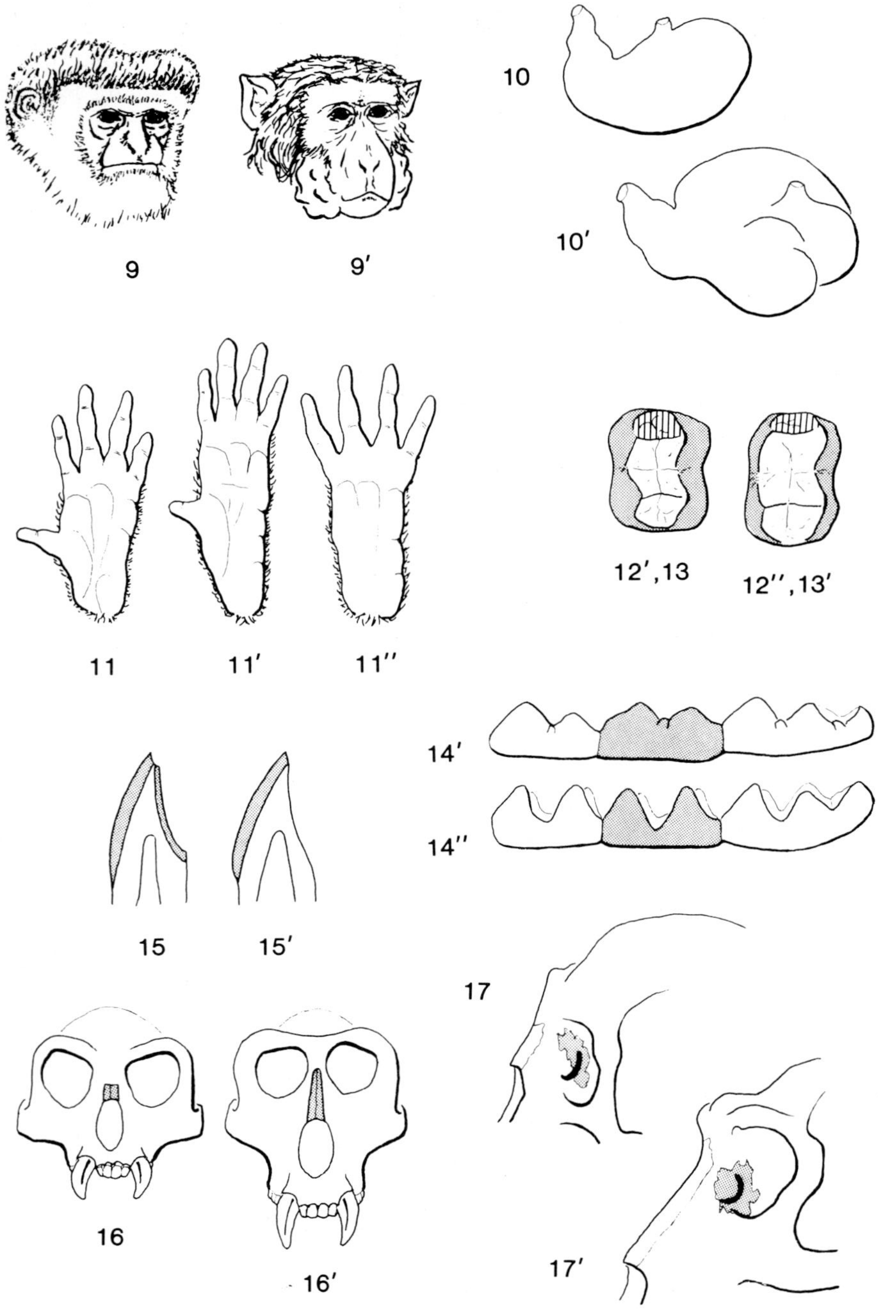

Figure 4. Selected characters at Node 2. Anterior view of face: 9 buccal pouches absent, 9′ buccal pouches present. Anterior view stomach, pyloric region to left: 10 stomach simple, 10′ stomach multi-chambered. Palmar view left hand: 11 pollex (thumb) well developed, 11′ pollex reduced, 11″ pollex absent. Dorsal view RM$_2$, mesial to top, buccal to right: 12 cingulum present (not illustrated), 12′ molar flare well developed (stippled), 12″ molar flare secondarily reduced; 13 trigonid long (parallel lines), 13′ trigonid short. Lingual view RM$_{1-3}$, mesial to left: 14 relief low, 14′ relief high. Longitudinal cross-section I$_1$, lingual to right: 15 enamel (stippled) present lingually, 15′ lingual enamel absent. Anterior view skull: 16 nasal bones (stippled) short, 16′ nasal bones long. Oblique left view skull, anterior to left: 17 lacrimal bone (stippled) within orbit & lacrimal fossa (bold line) straddles maxillary and lacrimal bones, 17′ lacrimal bone at edge of orbit completely encloses lacrimal fossa.

outgroup. Both conditions in this postulated transformation series are found in *Propliopithecus*, where the molars of *P. markgrafi* do not show the cingulum present in those of its congenerics, but instead are flared, although not as much as are cercopithecid molars (Szalay & Delson, 1979).

The third feature in this complex, relative trigonid length, also is subject to alternative interpretation. Delson (1975) considered the colobines to have a relatively shorter trigonid basin than cercopithecines, and this observation has recently been supported by the quantitative work of Benefit (1987). Delson argued that parapithecids and propliopithecids have a rather long trigonid, implying that the cercopithecine condition was conservative and the colobine trigonid derived. Harrison (1982) and Benefit (1987) have countered that to them the trigonid appeared shorter in the eucatarrhine morphotype, implying that the colobine condition was less derived among cercopithecids. Resolution of this conflict requires quantitative and more detailed comparative analysis of this feature across anthropoids, which has not yet been undertaken.

The outcome may have a broader phylogenetic import, as Delson (1973) further argued that the presence in *Victoriapithecus macinnesi* of a short trigonid (as well as low relief) indicated that at least one derived colobine feature had already appeared in the middle Miocene. Moreover, he isolated at least one dental specimen (KNM MB 34) and a damaged distal humerus (KNM MB 19) as indicative of a potential second taxon in the Maboko Island assemblage and suggested that the cercopithecine–colobine divergence might be represented there, at 15 Ma (million years ago). Both Harrison (1982) and Benefit & Pickford (1986) have rejected this interpretation, arguing that only one cercopithecid species is present, but Senut (1986) has also recognized two taxa on postcranial evidence. Benefit's analysis of the greatly increased sample of Maboko monkeys convincingly demonstrates that only one taxon is common; MB 34 could represent a rare second species but is probably best accepted as a slightly distinctive individual of *V. macinnesi*. The postcranial differences among far fewer specimens require further analysis, but we here abandon Delson's (1975) hypothesis of two cercopithecid taxa at Maboko. If Benefit is correct about short trigonids being ancestral for cercopithecids, then *Victoriapithecus* shares few derived states with either modern subfamily, but we do not yet accept her interpretation. Butler (1986) further noted that in the character complex relating hypoconulid loss to molar wear orientations, *Victoriapithecus* presents a state intermediate between that of later cercopithecids (no cusp, phase II facets oriented transversely) and other eucatarrhines (cusp present, facets oblique). In *Victoriapithecus*, the hypoconulid has been lost (or is rarely present but small: Benefit, 1987), but the facets are oblique. The slightly older and much rarer *Prohylobates* is not yet known from unworn specimens in which the state is unambiguous, but there is a question as to whether it usually retains a hypoconulid (see Leakey, 1985).

To turn to the anterior dentition the distribution of enamel on the lower incisors in the cercopithecid morphotype was primitively as in most other primates, with a thick layer on the labial side of the teeth and a somewhat thinner deposit on the lingual. It was first noted by Noble (1969) and later confirmed by Delson (1975) that papionins, like rodents and New World marmosets (Rosenberger, 1978), have eliminated the lingual layer of enamel, effectively producing a self-sharpening, chisel-like edge on these teeth. Shellis & Hiiemae (1986) recently tested this observation using histological techniques and documented that this is a characteristic shared by all cercopithecines (Figure 4, char. 15). They furthermore argued that the blunt-edged colobine incisors are better designed for gripping and tearing leaves while the cercopithecine condition enhances the cutting and scraping that is often necessary for preparing fruits for consumption.

As regards facial architecture, Vogel (1968) postulated that the ancestral eucatarrhine had a broad interorbital distance with short nasal bones and a lacrimal fossa which straddled the border of the lacrimal and maxillary bones, as seen in colobines and most hominoids (Figure 4, char. 16–17; char. 23). Cercopithecines show a derived condition for these characters, all of which are probably associated with lengthening of the face. Thus, they have a narrower interorbital distance, longer nasal bones and a lacrimal fossa which lies entirely within a lacrimal bone that has expanded to such an extent that it extends outside the orbit. Additionally, cercopithecines have a longer and lower-vaulted cranium, a derived sulcul pattern, and a mandibular corpus which increases in depth mesially while the colobine morphotype retains the primitive high-vaulted cranium and cerebral sulci as well as a mandibular corpus which is expanded in the gonial region and shallows mesially (chars 24–26).

In the postcranium there are a number of derived features which characterize the colobine morphotype. These are not as readily correlated with their dietary specialization, but may be interpretable *vis-à-vis* their locomotor behavior. Colobines are agile leapers and runners in the forest canopy and generally are more slender in build than cercopithecines of the same body size, as is reflected in their longer limbs and iridia. It is especially in the latter that they have diverged from the cercopithecid morphotype. The most obvious synapomorphy is the reduction of the external pollex and, in some cases, its entire loss (Figure 4, char. 11). Furthermore, as Gabis (1960) noted, colobines have a relatively shorter tarsus for any given foot length than do cercopithecines and the inferred cercopithecid morphotype. Such shortening effectively reduces the lever arm for the extrinsic muscles working on the distal foot and is clearly reflected in the contacts of the cuboid with the ectocuneiform (Figure 5, char. 18). The primitive condition among anthropoids, and retained in cercopithecines, is to have a long proximal and short distal contact between these two bones. In colobines these proportions are reversed, and in some taxa the proximal contact is lost altogether.

In addition, colobines have modified the functional axis of the foot, that is, the line of weight bearing through the digit rays. It is inferred from the distribution of relative metatarsal length, robusticity and attachment of muscles that the axis primitively passes through the middle of the foot travelling along the third digit ray. In cercopithecines this primitive axis is retained and the external flexors of the digits, which pass into the foot behind the astragalus, are arranged so that digits 1, 3 and 4 are operated by *m. flexor fibularis* (*m. flexor hallucis longus*) and the medially disposed *m. flexor tibialis* (*m. flexor* digitorum longus) manages digits 1, 2 and 5. Colobines are inferred to have shifted the axis laterally, between digits 3 and 4, since the fourth digit is invariably equal in length and robusticity to the third (Figure 5, char. 20). Furthermore, in colobinans and some presbytinans, *m. flexor fibularis* completely loses its attachment to the hallux while in most presbytinans its contribution appears to be reduced (char. 27; Polak, 1908; Lewis, 1962). Thus, at least in many colobines, *m. flexor fibularis* is solely responsible for managing the elongated digits which flank the functional axis. This leaves *m. flexor tibialis* as the exclusive hallucal operator, albeit with assistance from *m. flexor accessorius* (*m. quadratus plantae*). This probably explains the presence of a groove on the proximal astragalus in colobines which supports a well developed retinaculum that in turn resists the displacement of *m. flexor tibialis* during its activity (Figure 5, char. 19).

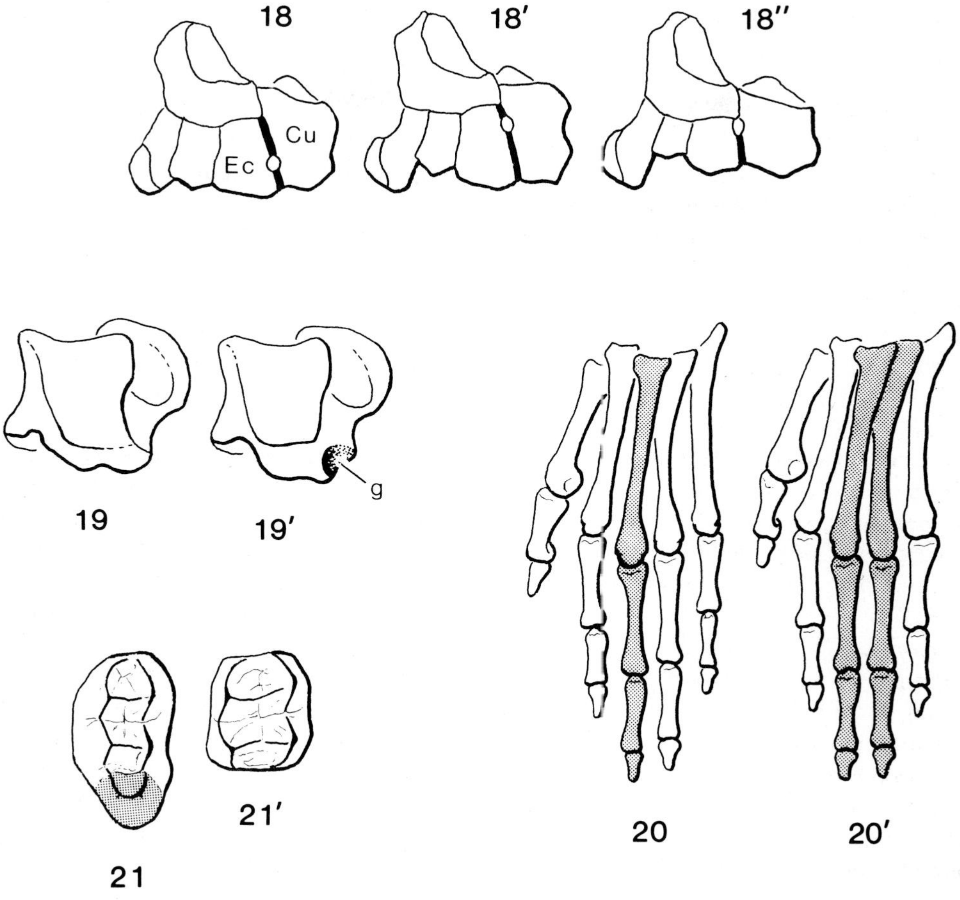

Figure 5. Selected characters at Nodes 2–4. Dorsal view left distal tarsals, proximal to top, lateral to right, Ec ectocuneiform, Cu cuboid, bold lines indicate articulations: 18 proximal facet long & distal facet short, 18′ proximal facet short & distal facet long, 18″ proximal articulation missing. Proximal view left astragalus, lateral to left: 19 m. flexor tibialis groove (g) absent, 19′ groove well developed. Dorsal view left pedal digit rays, medial to left: 20 functional axis through third digit ray (stippled), 20′ functional axis between digit rays 3–4. Dorsal view RM_3, mesial to top, buccal to right: 21 hypoconulid (stippled) present, 21′ hypoconulid absent.

Major subdivisions within the extant subfamilies

The Cercopithecinae are today commonly divided into two tribes (following Kuhn, 1967): Papionini, the macaques, baboons, drills, mangabeys and gelada; and Cercopithecini, the forest guenons and relatives. The papionins have secondarily increased the maximum width of their nasal aperture (Figure 3, char. 5) and generally have both relatively longer faces and stronger adaptations to terrestrial life than does the cercopithecine morphotype (chars 28–29). Cercopithecins, on the other hand, show three synapomorphies compared to the ancestral condition. They have lost the hypoconulid on the third molar, and in parallel with colobines, all cercopithecins except *Allenopithecus* have secondarily reduced the amount of molar flare (Figure 4, char. 12; Figure 5, char. 21). Finally, cercopithecins show a great deal of fragmentation of their chromosomes, having diploid numbers between 48 and 72, while papionins all have 42 and colobines mainly 44 (char. 30).

The subfamily Colobinae was tentatively divided by Szalay & Delson (1979) into two groups which appear to be less distinct than the cercopithecine tribes and which were therefore ranked as subtribes: the Asian Presbytina and the African Colobina. As noted above, characters 11 and 18 presented a complex morphocline, and it is the presbytinan morphotype which retains the more primitive colobine condition for both characters. This comprises a reduced external pollex and a small proximal contact between the lateral tarsals, while the ancestor of African colobinans was more derived in eliminating both of these features.

Notes on generic relationships within higher taxa

The majority of the foregoing analysis is well-accepted, with only the potential reversal in character 13 (trigonid length) leading to hesitation. The greatest uncertainty relates to generic relationships within each of the four cercopithecid taxa reviewed above. Within Papionini, Szalay & Delson (1979) recognized three subtribes as did Jolly (1966) and Kuhn (1967). On mainly biogeographic grounds Delson (1973, 1975) postulated that the origin of these subtribes was related to their separation by the enlargement of the Sahara desert, interpreted as a vicariant event in the late Miocene (*ca* 9–6 Ma). Neither the North African and Eurasian Macacina nor the sub-Saharan Papionina are characterized by consistent derived features, while the monotypic Theropithecina is simply autapomorphic. Living and fossil *Macaca* can be separated from African papionins as a genus which persistently retains primitive states, especially in terms of facial profile, lack of facial fossae and locomotor morphology (chars 29, 31–32). Two extinct large-bodied Eurasian genera appear to be "derived macaques", differing mainly in their greater adaptations to terrestriality. Until *Procynocephalus* (late Pliocene of India and China) is better-known cranially, and it is possible to determine if it shared the apparently reduced cranial sexual dimorphism seen in *Paradolichopithecus* (Pliocene of Eurasia), it is perhaps best to retain these similar forms as distinct genera. Compared to the Macacina, most papioninans including *Theropithecus*, share greater facial elongation with a steep anteorbital drop, development of facial fossae and increased adaptations to terrestriality. Within this group, *Theropithecus* is clearly the most derived dentally, facially and postcranially (see Jolly, 1972). It is first documented in the fossil record at least 4 Ma, and on the assumption that a considerable period of time was required to develop its numerous autapomorphies, Szalay & Delson (1979) recognized a subtribe Theropithecina. What remains uncertain is whether *Theropithecus* is the sister-taxon to all (other) papioninans or a member of that taxon, with its closest relative one specific genus (such as *Papio*). Despite its early presence in the fossil record, *Theropithecus* could well have been a rapidly evolved offshoot from a *Papio*-like stock, as Cronin & Meikle (1979) suggested partly on the basis of molecular studies. At present, it would seem best to include *Theropithecus* within Papionina, whose other relationships also require further analysis. For example, the monophyly of the mangabeys needs detailed investigation (Groves, 1978; Kluberdanz, in prep.), as does the placement of *Mandrillus* as either a subgenus of *Papio* or a quite distinct genus. It may be thought inconsistent to retain *Paradolichopithecus* while eliminating Theropithecina, but the combination of the two Eurasian genera (if required by future analysis) is less confusing than the erection of an infrasubtribal category and taxon for *Theropithecus* which now appears to belong within the Papionina. Of the Plio-Pleistocene fossil papionins, *Gorgopithecus* (late Pliocene of S. Africa) is clearly derived in its reduced dental sexual dimorphism and mid-facial/zygomatic morphology, but the contemporaneous *Dinopithecus* (also known from the Pliocene of Ethiopia and Angola) is best considered as a subgenus of *Papio* (see Delson, 1984). The long-lived *Parapapio* (late Miocene to late Pliocene of Africa) is conservative, in many ways similar to *Macaca*, although its facial profile is straighter and less concave, possibly implying a transformation series whose polarity is yet unclear. Until the relationships of the living genera are better clarified, it is nearly impossible to interpret these still fragmentary fossil finds.

The Cercopithecini is a less diverse group at the generic level, but the relationships of the very successful *Cercopithecus* species continue to frustrate attempts at resolution (cf. Maranto, 1986 and Muleris *et al.*, 1986). Ignoring that problem leaves four species, often placed in three distinct genera: *Allenopithecus, Erythrocebus* and *Miopithecus*. Of these, the first would seem to be readily interpreted as a species with many retained ancestral cercopithecine features such as molar flare, continuous male ischial callosities and cyclic changes in female sexual skin; it also has only 48 chromosomes, the lowest diploid number in the tribe (chars 12, 30, 33–34). The difference between *Allenopithecus* and the remaining cercopithecins appears roughly as distinct as that between the subtribes of Papionini and Colobinae. On this basis we suggest formal recognition of a subtribe Allenopithecina, defined by the three cited retentions and based on *A. nigroviridis*.

It may be argued that one should not define a taxon based solely on conservative retention of characters which define the higher taxon to which it belongs. However, there are three points to consider in this regard. First, it is quite probable that further study may permit the delineation of autapomorphies for *Allenopithecus*. Moreover, this species does present the derived characters of the Cercopithecini, although it is conservative by comparison to other species in that tribe. Most importantly, if one insists that any taxon must be defined by autapomorphies, and at the same time argues that the presence of an autapomorphy prevents one species from being ancestral to its sister taxon, the possibility of ever identifying ancestors is excluded. Ancestral species obviously did exist, and if a "real" ancestor cannot be recognized because of a definitional stumbling block, systematic progress is surely not being furthered. A similar problem can be recognized with regard to *Homo erectus*, a species probably ancestral to the broadly defined *Homo sapiens*, but showing autapomorphies only in Asian populations (see Rightmire, 1987). This is not to say that *Allenopithecus* is being suggested as the potential ancestor of the Cercopithecina—what we are addressing here is the broader methodological argument that non-autapomorphic taxa may have a place in systematics without running the risk of being dismissed as paraphyletic (in the definition, for example, of Patterson, 1982).

The nominate cercopithecin subtribe, Cercopithecina Gray, 1825, would be defined by the derived loss of molar flare and connected male ischial callosities, as well as an increased diploid number. Among cercopithecinans, only *Miopithecus* presents sexual skin cyclicity, and on this basis Groves (1978) has questioned its placement even among the Cercopithecini. This appears unlikely in light of the presence in *Miopithecus* of all cercopithecin and three of the four cercopithecinan characters. Moreover, Verheyen's (1962) analysis of cranial form, among others, convinces us that *Miopithecus* is a neotenous, phyletically dwarfed *Cercopithecus*, but the presence of female cycling leads us to rank it as a full genus, perhaps the most distinct one in the subtribe. *Erythrocebus*, on the other hand, is more distinct cranially and postcranially, but has lost cyclicity. Both of these taxa have low diploid numbers of 54, suggesting fairly early differentiation from the main *Cercopithecus* lineage, but they have not yet been unambiguously linked to any species groups thereof.

The phyletic geometry of the Colobinae is even more confused, with the two subtribes being neither widely accepted nor definitely holophyletic. Among the Colobina, three extant taxa are commonly recognized, often as the single genus *Colobus*, with three subgenera. However, some recent classifications, such as those by Thorington & Groves (1970) and Napier (1985) following Groves (1970), link *Piliocolobus* with *Colobus*; others, especially Kuhn (1967), include *Piliocolobus* in *Procolobus*. Napier (1985) has clearly tabulated a series of characters which vary among these three species groups, consistently grouping *Piliocolobus* and *Procolobus*. The polarity of some of these characters is uncertain, but it appears to us that the lack of female sexual swellings (also found in presbytinans), and the probably inter-related large larynx and sub-hyoid sac are derived features of *Colobus*, while the discontinuous male ischial callosities (also seen in *Pygathrix*), male perineal organ and four-chambered stomach are synapomorphies of *Procolobus* and *Piliocolobus* (chars 10, 34–35; Kuhn, 1972). Brandon-Jones (1985) has grouped these two taxa in *Procolobus*, adding the common presence of a sagittal crest in males; we tentatively consider this feature also a true synapomorphy, possibly implying relatively smaller brain size. Pending further work in progress, we recognize *Procolobus* to include subgenera for the olive and red colobus, with *Colobus* restricted to the black and white group. *P.* (*Procolobus*) is further distinguished by the (presumably conservative) lack of pattern in its pelage and by the presence of a median mental foramen, otherwise common in cercopithecines, rare in hominoids (10–20% occurrence) and very rare to absent in other extant colobines (char. 36; Vogel, 1968). Yet, the Plio-Pleistocene African fossil colobines *Cercopithecoides* and *Rhinocolobus* also present this foramen, and its phyletic implication is thus uncertain (Leakey, 1982). These and other African fossils are tentatively included in the Colobina, although most are too fragmentary to permit determination of the diagnostic features of the subtribe: extreme pollicial reduction, the loss of the proximal contact between cuboid and ectocuneiform (the latter state is observable in the Pliocene *Paracolobus*), and loss of insertion by *m. flexor fibularis* onto the hallux.

There is perhaps even less agreement about the relationships among Asian colobines. We have presented presbytinans as an unresolved trichotomy in Figure 1 because there is no one character uniting any two of the three lesser groups. At the generic level, we follow Groves (1970) in including *Simias* within *Nasalis* and *Rhinopithecus* as a subgenus of *Pygathrix*, Delson (1975) in allowing a subgenus for *N.* (*Simias*) and tentatively Brandon-Jones (1985) in distinguishing *Semnopithecus* and *S.* (*Trachypithecus*) from *Presbytis*. Groves (1970) has made the most detailed analysis of this group, but could find no clearly derived features which linked any pair of genera. Even the long-accepted grouping of *Presbytis* with *Semnopithecus* (s.l.) is based on no more than a mainly intermediate condition in facial structure (e.g., face shorter than *Nasalis* but longer than *Pygathrix*), relatively shorter arm compared to leg length (char. 37; like most colobinans, but not the apparently conservative *Nasalis* and *Pygathrix*), and other features which are probably of low phyletic value. Kuhn (1967) went so far as to suggest that *Nasalis* might be the sister-taxon to all other colobines, a view which cannot be fully refuted as yet.

The relationship between Asian and African colobines is thus problematical. Other than geographic separation, a few features clearly distinguish members of the two subtribes. Extant colobinans certainly form a clade whose ancestor was characterized by derived thumb reduction and tarsal shortening. However, both of these features are parts of trends seen in Presbytina which, as mentioned above, may be a paraphyletic taxon. Further study is required to discern the more precise relationships of individual Asian colobines to the African forms. Consideration of the extinct Mio-Pliocene European colobines *Mesopithecus* and *Dolichopithecus* leads to further confusion. The younger genus, *Dolichopithecus*, was thought to be more derived on a number of morphoclines than its purported ancestor, representing one of the few acceptable cases of a generic ancestor-descendant relationship in the catarrhine fossil record (Delson, 1975). More recent studies (Strasser, in prep. *a*) reveal that while *Mesopithecus* presents the derived loss of the proximal cuboid-ectocuneiform facet, *Dolichopithecus* retains that contact, albeit reduced, much as in *Pygathrix* and *Nasalis*. Such loss in *Mesopithecus* and retention in *Dolichopithecus* weakens the hypothesis of direct descent in the European clade but does not imply any phyletic relationship for these forms to extant colobines. Indeed, until the phylogeny of Asian colobines is better resolved the taxonomic allocation of these European forms must remain Colobinae *incertae sedis*.

The resulting classification of Cercopithecidae to the level of subgenus is presented in Table 3. We have discussed in the text most modifications to existing classifications. However, a brief comment on the phyletic and taxonomic position of *Victoriapithecus* (and *Prohylobates*) may be worthwhile. These mid-Miocene species are especially interesting because they are the oldest representatives of the Old World monkeys. Benefit (1987) has argued that they should be grouped in a monophyletic taxon (named as Victoriapithecinae by Koenigswald, 1969) on the basis of a shared derived mandibular symphysis morphology and the supposedly oblique orientation of the long axis of P_4 (see also Leakey, 1985, who implied a possible synonymy of the two genera). Although many characters are unknown in these fossils, and the polarity of others (such as the trigonid length) is uncertain, they may not share any derived conditions with either later subfamily. On this basis, Benefit (1987) has proposed that the mid-Miocene taxa be placed in a distinct family of Cercopithecoidea, but we prefer to retain them as a subfamily until these arguments are further elaborated and the character polarities assessed, demonstrating our uncertainty with a trichotomy.

Summary

In this paper we present the derived features which characterize the hypothetical ancestors of Cercopithecidae and its subclades. While some typical features have been recognized since the time of Linnaeus, continued research reveals additional characteristics that allow the refinement of hypotheses of phylogenetic relationship. The cercopithecid ancestor is distinguished by 13 derived features which we suggest allowed it to successfully exploit a "niche" unavailable to early hominoids. These include: the alignment of cusps into a bilophodont condition; the incorporation of cingulum into the molar sidewall resulting in a flared tooth; the elongation and increased relief of molars; the loss of the hypoconulid on dP_4–M_2; the elongation of the P_3 flange for increased honing of C^1; the extension of C^1 mesial sulcus onto the root; the development of a high and narrow nasal aperture; the modification of the posterior and anterior compartments of the lower ankle joint to stabilize it and limit the extent of pedal inversion; and the reduction of hallucial axial rotation during ab-adduction. The mid-Miocene *Victoriapithecus* and *Prohylobates* appear to be linked by one possibly derived feature, but this clade is ranked as a subfamily rather than a

Table 3 **Outline classification of Cercopithecidae**

- Order Primates
  - Semiorder Euprimates
    - Suborder Haplorhini
      - Hyporder Anthropoidea
        - Infraorder Catarrhini
          - Parvorder Eucatarrhini
            - Superfamily Cercopithecoidea
              - Family Cercopithecidae
                - Subfamily Cercopithecinae
                  - Tribe Cercopithecini
                    - Subtribe Allenopithecina
                      - *Allenopithecus*
                    - Subtribe Cercopithecina
                      - *Cercopithecus*
                      - *Miopithecus*
                      - *Erythrocebus*
                  - Tribe Papionini
                    - Subtribe Papionina
                      - *Papio*
                      - *P. (Mandrillus)*
                      - †*P. (Dinopithecus)*
                      - *Cercocebus*
                      - *"Lophocebus"*
                      - †*Gorgopithecus*
                      - *Theropithecus*
                      - †*Parapapio*
                    - Subtribe Macacina
                      - *Macaca*
                      - †*Procynocephalus*
                      - †*Paradolichopithecus*
                - Subfamily Colobinae
                    - Subtribe Colobina
                      - *Colobus*
                      - *Procolobus*
                      - *P. (Piliocolobus)*
                      - †*Libypithecus*
                      - †*Cercopithecoides*
                      - †*Paracolobus*
                      - †*Rhinocolobus*
                    - Subtribe Presbytina
                      - *Presbytis*
                      - *Semnopithecus*
                      - *S. (Trachypithecus)*
                      - *Pygathrix*
                      - *P. (Rhinopithecus)*
                      - *Nasalis*
                      - *N. (Simias)*
                - Subfamily Colobinae, incertae sedis
                      - †*Mesopithecus*
                      - †*Dolichopithecus*
                - Subfamily Victoriapithecinae
                      - †*Victoriapithecus*
                      - †*Prohylobates*

† Extinct genus or subgenus.

separate family to emphasize our uncertainty about its phyletic relationship to the later subfamilies.

The cercopithecine morphotype is characterized by the acquisition of buccal pouches; loss of lingual enamel on the lower incisors; facial elongation with the concomitant decrease in interorbital distance, lengthening of nasal bones and complete enclosure of the lacrimal fossa within the lacrimal bone; elongation of a low-vaulted cranium; development of a derived cerebral sulcal pattern and a mandibular corpus which increases in depth mesially. Within Cercopithecinae there are two tribes, each one of which contains one persistently conservative and one holophyletic and probably derived subtribe. The morphotype for the tribe Papionini shows a secondary increase in the maximum width of their nasal aperture; facial elongation; and stronger adaptations to terrestrial life. Macacinans retain the morphotypic conditions while papioninans show a trend in developing: a steep anteorbital drop; facial fossae; and further postcranial adaptations to terrestrial life. *Theropithecus* is included within the latter subtribe until further study clarifies the relations of fossil and recent baboons and mangabeys. The morphotype for the tribe Cercopithecini is distinctive in the loss of the hypoconulid on M_3 and the tendency towards chromosomal fragmentation. Within this tribe, the morphotype for Cercopithecina (containing *Cercopithecus, Miopithecus* and *Erythrocebus*) is further derived by secondarily reducing the amount of molar flare and separating male ischial callosities. Guenons and patas, but not talapoins, have also lost cyclic changes in female sexual skin, which leads us to rank *Miopithecus* as a genus, rather than the subgenus suggested elsewhere. The newly recognized subtribe Allenopithecina retains the more conservative condition for these features. It is reasonable to recognize such "persistently primitive" taxa both because they do present derived features of their clade as a whole and in order to avoid rendering the identification of putative ancestral species definitionally impossible.

The colobine morphotype presents: a complex, sacculated stomach; molar teeth in which flare is secondarily reduced and the trigonid shortened; a mandible with an expanded gonial region; a reduction of the pollex; a shifting of the functional axis of the foot with the development of a proximal astragalar groove for *m. flexor tibialis*; and a shortened tarsus reflected in the contacts of the cuboid and ectocuneiform. Two subtribes are recognized within Colobinae: Presbytina for the Asian colobines and Colobina for the African. The presbytinan morphotype is the more primitive of the two, retaining a small external thumb and a reduced proximal facet between the cuboid and ectocuneiform. The colobinan morphotype is characterized by three losses: of the external pollex; the proximal facet between the cuboid and ectocuneiform; and the insertion of *m. flexor fibularis* onto the hallux. Within Colobina we recognize *Colobus* for the black-and-white colobus and *Procolobus* for the red and olive forms. The former is characterized by apparently derived enlargement of the larnyx and presence of a sub-hyoid sac. *Procolobus* and *P.* (*Piliocolobus*) share four synapomorphies: a four-chambered stomach, male perineal organ, discontinuous ischial callosities and a sagittal crest. Among the presbytinans, generic distinction is less clear, but at present we recognize *Nasalis, N.* (*Simias*), *P.* (*Pygathrix*), *P.* (*Rhinopithecus*), *Presbytis, S.* (*Semnopithecus*) and *S.* (*Trachypithecus*). While the phyletic relationships among fossil and living colobines are not yet resolved, pedal characters are beginning to clarify the situation; e.g., *Mesopithecus* presents a derived feature not seen in its putative descendant *Dolichopithecus*, casting doubt upon this hypothesis of descent.

Acknowledgements

We thank Fred Grine, John Fleagle and Lawrence Martin for inviting us to participate in the Hennig VI Symposium, for which this paper was prepared; Chester Tarka, Lorraine Meeker and Suzanne Walker for advice on the preparation of illustrations by E. S.; Guy Musser, David Dean and three reviewers for comments on drafts of the manuscript; and the curators and staffs of the following institutions for access to collections in their care: American Museum of Natural History, Museum of Comparative Zoology, National Museum of Natural History, Field Museum of Natural History, The University of Wisconsin-Milwaukee Tappen collection, British Museum (Natural History), Powell-Cotton Museum, Musée Royal de l'Afrique Centrale, Rijksmuseum van Natuurlijke Historie, Zoologisk Museum (Copenhagen), Zentrum Anatomie (Georg-August-Universität, Göttingen), Natur-Museum Senckenberg, Zentrum der Morphologie der Klinikum der Johann Wolfgang Goethe-Universität, Naturhistorisches Museum Wien, Naturhistorisches Museum Basel, Museo di Geologia e Paleontologia (Firenze), Anthropologisches Institut und Museum der Universität Zürich-Irchel, Muséum National d'Histoire Naturelle (Paris), the National Museums of Kenya, the Transvaal Museum, the Paleontological Museum of the Academy of Sciences (Moscow) and the Institute of Vertebrate Paleontology and Paleoanthropology (Beijing).

Research upon which this work is based was supported, in part, by PSC-CUNY grants 13610 & 667370 and NSF grant BNS-8419939 to E.D., NSF BNS-8407911 to E.D. and E.S., and a C.U.N.Y. Mina Rees Award, Smithsonian Predoctoral Fellowship and Wenner-Gren grant to E.S.

References

Andrews, P. (1981). Species diversity and diet in monkeys and apes during the Miocene. In (C. B. Stringer, Ed.) *Aspects of Human Evolution*, pp. 25–61. London: Taylor and Francis.

Andrews, P. (1985). Family group systematics and evolution among catarrhine primates. In (E. Delson, Ed.) *Ancestors: The Hard Evidence*, pp. 14–22. New York: Alan R. Liss.

Bauchop, T. & Martucci, R. W. (1968). Ruminant-like digestion of the langur monkey. *Science* **161**, 698–700.

Benefit, B. (1987). The molar morphology, natural history and phylogenetic position of the middle Miocene monkey *Victoriapithecus* and their implications for understanding the evolution of Old World monkeys. Ph. D. Dissertation, New York University.

Benefit, B. & Pickford, M. (1986). Miocene fossil cercopithecoids from Kenya. *Am. J. phys. Anthrop.* **69,** 441–464.

Brandon-Jones, D. (1985). *Colobus* and leaf-monkeys. In (D. MacDonald, Ed.), *The Encyclopedia of Mammals*, pp. 398–408. New York: Facts-on-File.

Butler, P. (1986). Problems of dental evolution in the higher primates. In (B. Wood, L. Martin & P. Andrews, Eds) *Major Topics in Primate and Human Evolution*, pp. 89–106. Cambridge: Cambridge University Press.

Cronin, J. E. & Meikle, W. E. (1979). The phyletic position of *Theropithecus:* Congruence among molecular, morphological and paleontological evidence. *Syst. Zool.* **28,** 259–269.

Delson, E. (1973). Fossil colobine monkeys of the circum-Mediterranean region and the evolutionary history of the Cercopithecidae (Primates, Mammalia). Ph.D. Dissertation, Columbia University, New York.

Delson, E. (1975). Evolutionary history of the Cercopithecidae. *Contr. Primat.* **5,** 167–217.

Delson, E. (1984). Cercopithecid biochronology of the African Plio-Pleistocene: correlation among eastern and southern hominid-bearing localities. *Cour. Forsch.-Inst. Sencken.* **69,** 199–218.

Gabis, R. (1960). Les os des membres des singes cynomorphes. *Mammalia* **24,** 577–602.

Gebo, D. L. & Simons, E. L. (1987). Morphology and locomotor adaptations of the foot in early Oligocene anthropoids. *Am. J. phys. Anthrop.* **74,** 83–101.

Groves, C. P. (1970). The forgotten leaf-eaters, and the phylogeny of the Colobinae. In (J. R. Napier & P. H. Napier, Eds.) *Old World Monkeys*, pp. 555–587. London: Academic Press.

Groves, C. P. (1978). Phylogenetic and population systematics of the mangabeys (Primates: Cercopithecoidea). *Primates* **19,** 1–34.

Harrison, T. (1982). Small-bodied apes from the Miocene of East Africa. Ph.D. Dissertation, University of London.

Harrison, T. (1987). The phylogenetic relationships of the early catarrhine primates: A review of the current evidence. *J. hum. Evol.* **16,** 41–80.

Jolly, C. J. (1966). Introduction to the cercopithecoidea, with notes on their use as laboratory animals. *Symp. Zool. Soc. Lond.* **17,** 427–457.

Jolly, C. J. (1972). The classification and natural history of *Theropithecus (Simopithecus)* (Andrews, 1916), baboons of the African Plio-Pleistocene. *Bull. Br. Mus. (nat. Hist.), Geol.* **22,** 1–123.

Kluberdanz, R. (in prep.). Are the mangabeys (Primates: *Cercocebus*) monophyletic? Ph.D. Dissertation, City University of New York.

Kuhn, H. J. (1964). Zur Kentniss des Bau und Funktion des Magens der Schlankaffen (Colobinae). *Fol. primat.* **2,** 193–221.

Kuhn, H. J. (1967). Zur systematik der Cercopithecidae. In (D. Starck, R. Schneider & H. J. Kuhn, Eds) *Neue Ergebnisse der Primatologie*, pp. 25–46. Stuttgart: G. Fischer.

Kuhn, H. J. (1972). On the perineal organ of male *Procolobus. J. hum. Evol.* **1,** 371–378.

Langdon, J. H. (1986). Functional morphology of the Miocene hominoid foot. *Contr. Primat.* **22,** 1–226.

Leakey, M. G. (1982). Extinct large colobines from the Plio-Peistocene of Africa. *Am. J. phys. Anthrop.* **58,** 153–172.

Leakey, M. G. (1985). Early Miocene cercopithecids from Buluk, northern Kenya. *Fol. primat.* **44,** 1–14.

Lewis, O. J. (1962). The comparative morphology of m. flexor accessorius and the associated long flexor tendons. *J. Anat.* **96,** 321–333.

Lewis, O. J. (1972). The evolution of the hallucial tarsometatarsal joint in the Anthropoidea. *Am. J. phys. Anthrop.* **37,** 13–34.

Maranto, G. (1986). Will guenons make a monkey of Darwin? *Discover* **X**(11), 87–101.

Muleris, M., Coutourier, J. & Dutrillaux, B. (1986). Phylogénie chromosomiques des Cercopithecoidea. *Mammalia* **50,** 38–52.

Murray, P. (1975). The role of cheek pouches in cercopithecine monkey adaptive strategy. In (R. Tuttle, Ed.) *Primate Functional Morphology and Evolution*, pp. 151–194. The Hague: Mouton.

Napier, P. H. (1981). Catalogue of Primates in the British Museum (Natural History) and elsewhere in the British Isles. Part II: Family Cercopithecidae, Subfamily Cercopithecinae. London: British Museum (Natural History).

Napier, P. H. (1985). Catalogue of Primates in the British Museum (Natural History) and elsewhere in the British Isles. Part III: Family Cercopithecidae, Subfamily Colobinae. London: British Museum (Natural History).

Noble, H. W. (1969). Comparative aspects of amelogenesis imperfecta. *Proc. r. Soc. Med.* **62,** 1295–1297.

Patterson, C. (1982). Morphological characters and homology. In (K. A. Joysey & A. E. Friday, Eds) *Problems of Phylogenetic Reconstruction* (Systematics Association Special Volume No. 21), pp. 21–74. London: Academic Press.

Polak, C. (1908). *Die Anatomie des Genus Colobus*. Amsterdam: Johannes Muller.

Rightmire, P. (1987). Species recognition and *Homo erectus. J. hum. Evol.* **15,** 823–826.

Rose, M. (1987). Further hominoid postcranial specimens from the late Miocene Nagri Formation of Pakistan. *J. hum. Evol.* **15,** 333–367.

Rosenberger, A. L. (1978). Loss of incisor enamel in marmosets. *J. Mammal.* **59,** 207–208.

Senut, B. (1986). Upperlimb postcranial elements of Miocene cercopithecoids from east Africa: Implications for function and taxonomy (abstr.). *Primate rep.* **14,** 87.

Shellis, R. P. & Hiiemae, K. M. (1986). Distribution of enamel on the incisors of Old World monkeys. *Am. J. phys. Anthrop.* **71,** 103–113.

Strasser, E. (in prep. *a*). Pedal morphology of Cercopithecidae and the evolution of the foot in Catarrhini. Ph. D. Dissertation, City University of New York.

Strasser, E. (in prep. *b*) Pedal evidence for the origin and diversification of cercopithecid clades.

Swindler, D. R. & Erwin, J. (Eds) (1986). *Comparative Primate Biology*. Vol. 1, *Systematics, Evolution and Anatomy*. New York: Alan R. Liss.

Swofford, D. L. (1985). *PAUP: Phylogenetic Analysis Using Parsimony, version 2.4.* Champaign, Ill.: Illinois Natural History Survey.

Szalay, F. S. (1975). Haplorhine relationships and the status of the Anthropoidea. In (R. Tuttle, Ed.) *Primate Functional Morphology and Evolution,* pp. 3–22. The Hague: Mouton.

Szalay, F. S. & Delson, E. (1979). *Evolutionary History of the Primates*. New York: Academic Press.

Thorington, R. W. & Groves, C. P. (1970). An annotated classification of the Cercopithecoidea. In (J. R. Napier & P. H. Napier, Eds.) *Old World Monkeys,* pp. 631–647. London: Academic Press.

Verheyen, W. N. (1962). Contribution á la craniologie comparée des primates. *Mus. r. Afr. Centr. (Tervuren), Ann.,* sér in 8vo, Sci. Zool., No. 105, 1–247.

Vogel, C. (1968). The phylogenetical evolution of some characters and some morphological trends in the evaluation of the skull in catarrhine primates. In (B. Chiarelli, Ed.) *Taxonomy and Phylogeny of Old World Primates with References to the Origin of Man,* pp. 21–55. Turin: Rosenberg & Sellier.

Peter Andrews

Department of Palaeontology, British Museum (Natural History), Cromwell Road, London SW7 5BD, England

Lawrence Martin

Department of Anthropology, State University of New York, Stony Brook, N.Y. 11794, U.S.A.

Received 29 May 1987
Revision received and accepted 20 July 1987

Publication date November 1987

Keywords: cladistics, hominoids, molecular evolution, *Proconsul, Kenyapithecus, Dryopithecus, Afropithecus, Heliopithecus, Sivapithecus.*

Cladistic relationships of extant and fossil hominoids

There is general agreement that the hominoid primates form a monophyletic group, that the extant great apes and humans form a second clade within that group with the gibbons as the sister group, and that the African apes and humans form a third clade. Although it has recently been proposed that humans and orang utans are sister taxa and also that the great apes form a clade to the exclusion of humans, our analysis, particularly of the molecular evidence, supports the existence of an African ape and human clade. The major problem in hominoid phylogeny at present is the relationships of the species within this clade: morphological data generally support the existence of an African ape clade which is the sister group to humans; some molecular data also support this conclusion, but most molecular evidence indicates the existence of a chimpanzee/human clade. We have cladistically re-analysed the DNA and protein sequence data for which apomorphic character states can be assessed. It is clear that there is a high degree of homoplasy whichever branching pattern is produced, with some characters supporting the existence of a chimpanzee/human clade and others supporting an African ape clade. When the cladistic analyses of morphological and molecular data are combined we believe that the most parsimonious interpretation of the data is that the African apes form a clade which is the sister taxon of the human (i.e., *Australopithecus, Homo* and *Paranthropus*) clade.

This paper is not intended as a survey of all hominoid fossils but as a study of branching points in hominoid evolution and fossils are included which are relevant to this branching pattern. The analysis of fossil taxa in this study leads us to conclude that *Proconsul* is the sister taxon to the later Hominoidea. A number of middle Miocene forms such as *Dryopithecus, Kenyapithecus, Heliopithecus* and *Afropithecus* are shown to share derived characters with great apes and humans and provide evidence for the divergence of that clade from the gibbon lineage prior to 18 Ma. The position that *Sivapithecus* represents the sister group of the orang utan clade is supported here and shows that the orang utan lineage had diverged from the African ape and human lineage prior to 11·5 Ma. There is unfortunately no definitive fossil evidence on branching sequences within the African ape and human clade, although a new specimen from Samburu, Kenya may be related to the gorilla.

Journal of Human Evolution (1987) **16,** 101–118

Introduction

There have been a number of recent analyses of the relationships of the hominoid primates based on both morphological and molecular evidence. The two data sets have been in general agreement in terms of the recognition of a monophyletic group for Hominoidea and for a great ape and human clade. We have recently reviewed the evidence relating to these clades (Andrews, 1985, 1986, 1987; Martin, 1986), and a summary of some of this evidence is given in Figure 1 and will not be discussed further here. Instead, we will concentrate in this paper on the more controversial subject of the relationships within the great ape and human clade.

Molecular and most morphological studies have recognized the primary division of the orang utan from the African ape and human clade within the great ape and human clade. Schwartz (1984, 1986) has argued from morphological evidence that the orang utan is, in fact, the sister taxon to the human clade, and Kluge (1983) found evidence for a great ape

0047–2484/87/010101 + 18 $03.00/0

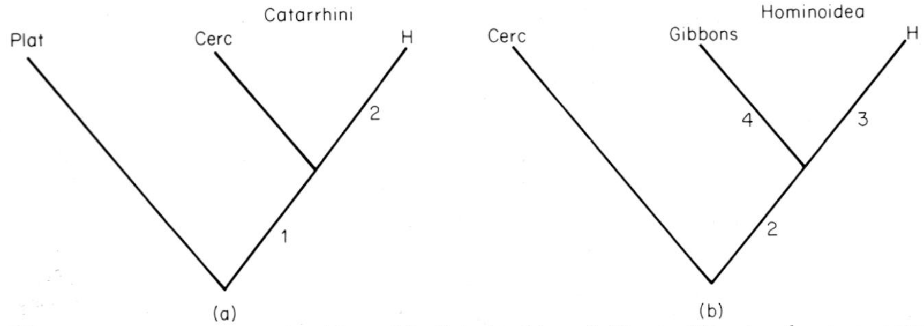

Figure 1. Hominoid relationships. (a) Relationship of Hominoidea to the two outgroups, Cercopithecoidea and Platyrrhini. (b) Initial branching point within Hominoidea between the gibbons and the great apes and humans. Synapomorphies for the four named branches are listed in detail in Andrews (1985) but we have selected out those characters that are particularly relevant to available fossil evidence:

Branch 1 Post-pliopithecid Catarrhine synapomorphies
skull with broad inter orbital pillar
broad glabella distinct from small supraorbital tori
reduced dental formula 2-1-2-3
lower P4 bicuspid
lower molars with hypoconulid, M3 largest
upper premolars heteromorphic
distal humerus with deep olecranon fossa, well defined lateral ridge, loss of entepicondylar foramen
talus longer than broad, long neck
development of a tubular ectotympanic

Branch 2 Hominoid synapomorphies
skull with enlarged paranasal sinuses
lower P3 low crowned
heteromorphy of upper premolars reduced
maxillary jugum developed
scapula with elongated vertebral border
humeral head rounded, larger than femoral head, more medially oriented
distal humerus with strong medial and lateral trochlear keels
reduced ulnar olecranon process
talus neck short, broad
loss of tail
reduction in lumbar, increase in sacral, vertebrae

Branch 3 Synapomorphies of the great apes and humans
skull with enlarged maxillary sinuses
orbits higher than broad
increased alveolar prognathism
lengthened premaxilla
abrupt transition from premaxilla to maxilla results in stepped floor of nasal chamber
incisive foramen small
lower P3 not bilaterally compressed, robust
enlarged premolars, upper premolars longer, lower premolars broader
lower P4 talonid equal in height to trigonid
upper premolar heteromorphy reduced
canines robust, reduced honing
upper molars with reduced cingulum
M2 largest molar, both upper and lower
molars with thick enamel
mandible robust, with deep symphysis
large inferior torus low on symphysis
distal humerus with deep sulci developed on either side of the lateral trochlear keel
reduced ulnar styloid process
molecular characters (see text)

Branch 4 Synapomorphies of the gibbons
numerous postcranial characters relating to brachiation listed in detail in Andrews & Groves (1975). No cranial or dental characters are known.

clade which excluded humans, but the evidence for these interpretations has recently been addressed by Andrews (1986, 1987) and Groves (1986) neither of whose analyses supported these claims. All molecular studies have consistently agreed in linking the African apes and humans, with the orang utan as the outgroup (Sarich & Wilson, 1967; Sarich & Cronin, 1976; Goodman, 1976; Romero-Herrera *et al.*, 1976, 1978; Brown *et al.*, 1982; Marks, 1983; Yunis & Prakash, 1982; Templeton, 1983; Sibley & Ahlquist, 1984). Despite disagreement about whether orang utans or African apes are most closely related to humans or whether the great apes form a clade which excludes humans, most morphological studies have recognized the existence of an African ape clade within the great ape and human clade (Kluge, 1983; Schwartz, 1984; Martin, 1986; Andrews, 1987). None of these authors have identified any specifically chimpanzee–human synapomorphies from gross morphology; on the contrary, they have emphasized the great morphological similarities shared by chimpanzees and gorillas. These have a very high probability of being shared derived characters because of their ontogeny and their complexity (Martin, 1985; Andrews, 1987). On the other hand, Groves (1986) has put forward some morphological evidence in support of a chimpanzee/human clade and this is backed up by a number of non-cladistic molecular studies, either based on overall similarity (Sibley & Ahlquist, 1984) or on parsimony analyses of sequence data (Goodman *et al.*, 1983; Hasegawa *et al.*, 1984, 1985; Koop *et al.*, 1986; Romero-Herrera *et al.*, 1978). Most recently, this conclusion has been supported by analyses of genetic distances from two-dimensional electrophoresis data, including a parsimony (PAUP) analysis of 240 electrophoretic characters (Goldman *et al.*, 1987). Other molecular biologists, however, have recognized an African ape clade to which humans are the outgroup (Brown *et al.*, 1982; Templeton, 1983).

This leaves unanswered three major questions in terms of extant hominoid relationships. These are as follows:

(1) Is there any evidence to support the position taken by Schwartz (1984) in recognizing an orang/human clade?
(2) Can the African apes be shown to form a clade separate from humans?
(3) If the answer to questions 1 and 2 is no, which of the African apes is most closely related to the human clade?

Methods of analysis

Fossils provide the primary source of historical data on hominoid evolution, but we feel that the fossil record is still too incomplete to base general statements about hominoid phylogeny on it alone. Consequently, we have undertaken firstly to analyse the cladistic relationships of extant forms, although we have emphasized those features which are likely to be relevant to the analysis of fossil taxa. On completion of the analysis for modern taxa, the fossils relevant to each branching point were evaluated as to their probable phylogenetic position on the basis of shared derived characters with recognized hominoid clades. We also accept that fossils add to the known morphological diversity of hominoids and have been prepared to modify our analysis of polarities obtained for the modern taxa in the light of contradictory fossil evidence where necessary.

The taxa included in this analysis were those conventionally recognized at the generic level and which constitute the superfamily Hominoidea, i.e. *Gorilla, Homo, Hylobates, Pan* and *Pongo*. For the purposes of this analysis we have identified monophyletic groups for the extant forms, with outgroups for Hominoidea using the extant Cercopithecoidea as the first outgroup and the Platyrrhini as a second outgroup. We have used two outgroups both as an added check that character states shared with the first outgroup are likely to represent the ancestral pattern, and to indicate a solution where commonality with the first outgroup fails to do so. By using two outgroups, it is possible to invoke parsimony as the grounds for a particular determination of polarity.

Hominoid synapomorphies have been identified where all hominoids show the same condition and differ from the cercopithecoid and the platyrrhine condition. In interpreting these as synapomorphies defining the Hominoidea they have played no further part in the analyses of relationships among the extant hominoids. The data base of characters used derives from Andrews (1985, 1986, 1987), Harrison (1982) and Martin (1985, 1986). For the molecular data we have developed the arguments made by Andrews (1986) so as to recognize those parts of the molecular data base which may be portioned into characters and analysed cladistically. As this part of our analysis is novel it is explained in greater detail below. Morphocline polarity has been determined on the basis of outgroup analysis and this has been considered in the light of the ontogenetic development of the character where possible (see Martin, 1985, 1986). We have also considered the functional nature of characters and character states as we believe that a knowledge of the function of a character is relevant to evaluating the likelihood of independent appearance through parallel evolution. Where sets of characters are considered to be functionally or developmentally correlated, e.g., those related to knuckle walking and those pertaining to enamel thickness and structure, we have treated these as multicharacter complexes (see discussion below).

All of the trees that can be produced from this kind of analysis involve some homoplasy and arguments have been made previously that minor differences of tree length should not be accorded undue weight. To preempt such problems we have placed greater emphasis on certain characters, i.e. we have weighted characters. We have given greater weight to characters on the grounds of three criteria all of which essentially relate to the degree of confidence with which the morphocline polarity can be determined and the extent to which we feel that homoplastic similarity can be eliminated as a possibility. In the first place, we have used simple characters whose polarity appears clear on the basis of outgroup analysis. Where both outgroups share a character state that is also found among some of the hominoid taxa, the polarity of character state change appears clear. Conversely, characters whose polarity could be interpreted in two ways with almost equal probability on the basis of outgroup comparisons are given lesser weight. We feel that this is essential, because otherwise a "robust" character can easily be outweighed by a number of labile characters whose polarity determination could almost be seen as the result of a flip of a coin. In fact, when one eliminates characters of dubious polarity from the initial analysis of relationships some of them may subsequently be seen as having potential for defining smaller clades. For example, a reduction of lumbar vertebral number to four may be seen as a derived character linking the two African apes if the common ancestor of the great ape and human clade is reconstructed as having five lumbar vertebrae and if the existence of a great ape clade is considered unlikely in the face of evidence from more robust characters. Naturally, the fact that reduction in lumbar vertebral number has occurred in two lineages, pongine and African ape, would suggest that this functionally important character can change in parallel relatively easily and should be accorded limited weight in any analysis. However, if other evidence exists that the African apes do form a clade then their shared reduction in lumbar vertebral number would be most parsimoniously interpreted as a retention from a common ancestor unique to them.

Secondly, any set of simple characters may also be given greater weight in the analysis if there are developmental data available for it. Even where the outgroup analysis does not provide a clear resolution of morphocline polarity, homology can be indicated when there is developmental evidence that supports one particular interpretation. This is indicated, for instance, in the developmental pattern of enamel accretion in the African great apes, where the enamel thickness appears primitively thin, but developmentally they share a method of achieving thin enamel which is a strong indication of shared homology and relationship. The orang utan also has secondarily reduced enamel thickness but this is the result of a different, step-wise, slowing down of secretion (Martin, 1983, 1985, 1986) and the secondary reduction of enamel thickness in the orang utan is therefore not only to a lesser extent than in the African apes but is not homologous developmentally.

Thirdly, where sets of characters are seen to be functionally or developmentally linked these may be treated as multicharacter complexes. When all of the elements of a multicharacter complex are distributed identically there is a very high degree of probability that the shared possession of the complex results from common ancestry. We recognize that it is possible that there may be a limited number of pathways available for the transformation of one multi-character complex state into another and that if such changes are under simple genetic control these multi-characters may be no better than single characters. However, this has yet to be shown to be the case and while accepting that they are not "better" characters, we believe that the use of such character complexes may help to reduce the problems of homoplasy in single characters, including, potentially, single nucleotide substitutions. This argument is elaborated below in reference to the molecular data, but has considerable value in treating morphological data (see Andrews, 1987). Particularly important for this analysis are the functional complex associated with knuckle walking in the African apes and the dental enamel complex in which enamel thickness, developmental rates and prism packing patterns are not only functionally linked but there is a direct developmental relationship among aspects of the complex whereby change in one character state results in change in another.

It has been argued that the shortest tree can rarely be obtained by inspection of even a moderate sized data matrix, as so many trees are possible and only computer algorithms can recognize the shortest one. We feel that both approaches have their merits but we have chosen to emphasize traditional cladistic character analysis rather than quantitative phylogenetics in this paper. In the approach that we have used, character polarity determination is the initial step and trees are then produced which most parsimoniously explain the distribution of derived characters among taxa. Many of the potentially possible trees are eliminated in the course of the polarity analysis. We would have no hesitation in using an algorithm to produce a tree from our data matrix once we have determined polarities, although this has not been done for the present paper, but we also feel that our own understanding of the nature of the characters might provide a better guide to selecting among trees of similar length.

Cladistic analysis of molecular data

Since 1963, when Morris Goodman published his phylogenetic analysis of serum proteins in primates, molecular evidence, in the broadest sense, has made major contributions to determining the branching sequence of hominoids as well as to the dating of such branching events even though the relative timing of these events is converted to absolute time by calibrating the clock from the fossil record. Many of these data sets produce distance statistics which have often been equated with phylogenetic proximity, a step which we feel requires justification. The molecular data for catarrhine phylogeny have been reviewed recently by Andrews (1986) and what follows is a development of his arguments. For the purposes of the present paper, we have been through the data reviewed by Andrews and selected those data sets that are amenable to cladistic analysis. Our criterion has been that character states must be definable so that an assessment of the polarity of character change can be made. Consequently we have only used sequence data in this discussion. This is not to say that distance data have nothing to contribute, simply that we remain to be convinced that they can be analysed in the cladistic fashion by which we are analysing our other data.

Amino acid sequence data

The evidence from this data set has been reviewed by Andrews (1986) and will not be discussed in any detail here. The African apes and humans share a secondarily derived character in having serine at position 23 of the myoglobin chain, and we recognize the possibility that at position 110 cysteine may define an African ape and human clade rather than the hominoid clade as suggested by Andrews (1986). In this case, the presence of cysteine in gibbon myoglobin at position 110 would be explained as the result of parallel evolution and the presence of serine in the orang utan as a primitive retention from the catarrhine ancestral condition. The substitution at position 23 is a robust character of clear polarity, but the one at position 110 is less certain.

In the haemoglobin sequence, the alpha chain at position 23 and the foetal globin A chain at position 104 have two amino acid residues unique to chimpanzee and human which most likely represent synapomorphies. No other pair of species within the hominoid clade shares any unique amino acid substitution (Goodman *et al.*, 1983; Scott *et al.*, 1984; Andrews, 1986).

The fibrinopeptide sequences A and B show the African apes and humans to be identical and to differ from the orang utan by two substitutions and from the gibbons by three or four (Doolittle *et al.*, 1971). Following the reasoning of Andrews (1986) these are interpreted as derived relative to the ancestral hominoid condition, defining an African ape and human clade by two substitutions, and a great ape and human clade by at least one additional substitution.

In the carbonic anhydrase sequence, the data from Tashian *et al.* (1976) reveal two shared derived substitutions linking *Pan* and *Homo* that are absent from *Pongo*. Unfortunately no data are available for *Gorilla*, so it is not possible to say whether this implies a chimp/human clade, or an African ape/human clade.

Restriction enzyme mapping produces ambiguous results, depending on the method of analysis (Ferris *et al.*, 1981; Bishop & Friday, 1986). On the other hand, it has been shown that chimpanzees and gorillas share a deletion in the ribosomal RNA chain (Hixson & Brown, 1986), and this appears to constitute an African ape synapomorphy.

Cladistic analysis of DNA sequence data

Two DNA sequences are available in the literature suitable for cladistic analysis. The method of analysis that we will use has been outlined briefly for a short sequence of mitochondrial DNA (mtDNA) (Andrews, 1986), and we propose to explain it more fully here for a longer sequence of nuclear DNA. Koop *et al.* (1986) have published the sequence of 2251 base pairs of the n-globin gene cluster. They formed genealogical trees by the most parsimonious placement of nucleotide substitutions. This procedure is based on adding all pairwise similarities in such a way as to minimize change, and it assumes that matches in nucleotide sequence are homologous, the result of shared common ancestry and not the result of convergence or back mutations. All similarities are given the same weight, and no account is taken of ancestral or descendent states.

For the first of these problems, the assumption of homology, there is little that can be done at the nuclear or any level. Individual nucleotide substitutions or single morphological characters appear the same whether they are homologous (result of common ancestry) or homoplastic (result of convergence, reversals, etc.), and there is no conclusive test that can distinguish them. This is a universal problem common to all methods of phylogenetic reconstruction, although perhaps especially of sequence and other simple data and less so for more complex systems as we have argued, and it is not the outcome of the application of any particular phylogenetic method, such as cladistics. One way around this problem is not to use single characters, whether morphological or molecular, but either to use groups of characters or characters for which some knowledge of ontogeny is available. Neither approach is presently possible for application to DNA sequence data, although complexity of character has been argued to justify the use of DNA–DNA hybridization (Sibley & Ahlquist, 1984). The bonding of DNA chains, and their resistance to dissociation, is the outcome of overall similarity of the chains, and while single base changes could easily be duplicated by homoplasy, the multiple group similarity, which provides the basis for overall similarity, could not easily be reproduced independently in two unrelated groups (Sibley & Ahlquist, 1984).

A similar argument for the use of multiple characters has also been raised with reference to morphological characters (Andrews, 1987). This states that if a number of characters are functionally correlated, the more complex the function and the greater the number of characters relating to it, the less likely it is to be achieved independently. The individual characters relating to the complex cannot be treated independently but must be regarded as part of a character complex, and for homology to be indicated all parts of this multiple character must be identical.

Illustrative example

For the second problem, the identification of ancestral or descendent states, cladistic methodology does offer a solution. It is claimed by many molecular biologists that their method is "cladistic", but the fact that it makes no effort to distinguish between ancestral and derived characters must make this claim of dubious value. To give an example of this, assume a distribution of four hypothetical character states of five taxa, as follows:

		Taxa				
		1	2	3	4	5
Character state	A	G	T	T	T	T
	B	G	G	T	T	T
	C	G	G	G	G	T
	D	G	G	G	G	G

This would give a matrix of similarity as follows:

	1	2	3	4	5
1	–	3	2	2	1
2		–	3	3	2
3			–	4	3
4				–	3

According to this analysis, the greatest similarity is shown by 3 and 4, but this is evidently based on a large amount of shared ancestral characters since for characters A, C and D all or nearly all the character states are the same in all five taxa. If now it is known, or assumed, that taxon 5 is the outgroup for taxa 1–4, the sharing of any character state between the outgroup and at least one of the related taxa indicates that this state is ancestral for the clade comprising taxa 1–4. Reworking the matrix of similarity counting only those shared character states that differ from the ancestral state produces the following matrix:

	1	2	3	4	5
1	–	2	1	1	0
2		–	1	1	0
3			–	1	0
4				–	0

Taxa 1 and 2 are now linked as sister taxa by this method, and all the other similarities are either the result of retained ancestral character states for taxa 1–4 which do not count, or must be considered homoplasies. The cladogram resulting from the character analysis is shown in Figure 2, which shows taxa 1 and 2 linked by character B, and this sister group linked equally with taxa 3 and 4 by character C. Character D, primitive in all five taxa, and character A, autapomorphic for taxon 1, do not contribute to our understanding of the relationships. In this instance, therefore, taking account of ancestral-descendent character

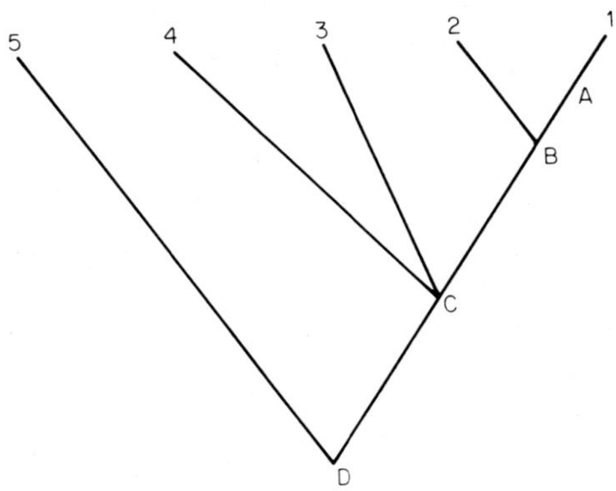

Figure 2. Cladogram resulting from hypothetical character analysis of five taxa, numbered 1 to 5, by four characters, A to D, each with just two character states, G and T. For the data matrix and explanation of the analysis, see the text. Position of characters A to D indicates the origin of the derived state linking taxa (above that node).

states radically alters our perception of relationship among these taxa by virtue of the removal from the analysis of shared primitive characters. Such a radical change may not always arise, but it is clear that failure to take account of this important distinction between different types of homology (shared ancestral or shared derived) always has the effect of including much redundant information in the form of shared retained ancestral characters.

This distinction can be applied to the sequence data provided by Koop *et al.* (1986). Comparisons of these data are based on the assumption that the New World monkeys (represented by the owl monkey in their analysis) are the outgroup for the catarrhine primates and the lemurs are the outgroup for the anthropoids, and from these we have deduced the ancestral condition for each position of the nucleotide sequence in relation to the hominoid clade (Figure 3) based on the following principles:

(1) All substitutions shared uniquely by any pair of species were counted, provided the other species, including the two outgroups, shared a single different substitution. In these cases it has been assumed that the condition common to the majority of species and the outgroup species represents the ancestral anthropoid condition from which the pair of species is derived. In our hypothetical example (above) taxa 1 and 2 share a single unique character state "G" in character B for which all other taxa, including the outgroup taxon 5, share character state "T". No other uniquely shared character state is present for any pair of taxa in this example. It has also been assumed that if one species pair differs from the others by a transversion rather than a transition, then a single transitional difference between the others is allowable (see below).

(2) All substitutions shared uniquely by any triplet of species were again counted provided the other species and the outgroups shared a single different substitution. Similarly for groups of four and five species, and again a single transition was accepted in the outgroup clades provided the sister clade differed by a transversion. In our hypothetical example, again, character C uniquely groups sister taxa 1 + 2 with 3 and 4 (Figure 2).

(3) Deletions and insertions were counted together and treated in the same way as substitutions. They were given equal weight, since they both represent single genetic events, but it is probable that the phenotypic consequences of the two types of change may be very different. Equal weight has not been given to the two types of nucleotide substitution: transitions (changes within purine or pyrimidine base pairs) are about ten times more likely than transversions (changes between pairs), so that the latter are much less common events (Brown *et al.*, 1982; Hasegawa *et al.*, 1984). Because of this high frequency of substitution, the chances of parallel change are thereby increased, and because also back mutations can occur with equal frequency, given the assumption of neutral effects of initial substitutions, it follows that transitions are more likely to be homoplastic than are transversions and are therefore less robust indicators of relationship than are transversions.

Applying these principles to the sequence data provided by Koop *et al.* (1986), the cladogram shown in Figure 3 is produced. Chimpanzees and gorillas share two genetic events, a deletion and a transition (G–A); chimpanzees and humans share three deletions and one insertion; and orang utans and humans share a single substitution (transition); no other genetic events are shared exclusively by any other pair of species. In all cases, the respective pair of species has one condition and the other primates all have the alternative

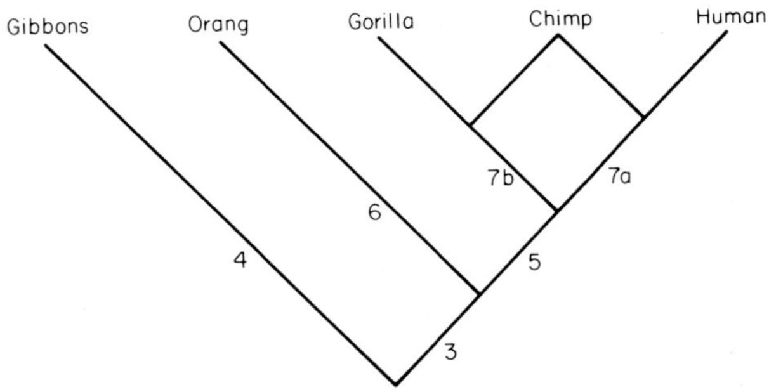

Figure 3. Branching points within Hominoidea. The numbering of the branches follows on from Figure 2, with branches 3 and 4 defined as before. Synapomorphies for the remaining branches are either discussed in the text or listed below. Data taken from Martin (1986), Groves (1986) and Andrews (1985, 1987).

Branch 5 Synapomorphies of the African ape and human clade
- development of supraorbital tori on the skull
- presence of frontal sinus developed ontogenetically by expansion from the ethmoid
- fusion of the os centrale
- shared substitution of amino acids in the myoglobin chain at positions 23 and perhaps 110
- two shared substitutions in the fibrinopeptide A and B chains
- nuclear DNA: 3 transversions
  8 transitions
  3 deletions
- mitochondrial DNA: 16 substitutions

Branch 6 Synapomorphies of the orang utan clade
- skull with loss of glabellar thickening
- interorbital distance narrow
- zygomaxillary foramen above lower rim of orbit
- facial profile concave
- extremely small incisive fossa caused by extensive overlap of premaxilla and maxilla
- smooth transition between premaxilla and maxilla
- massive, flattened zygomatic region
- great size discrepancy between upper I2 and I1

Branch 7a Synapomorphies of the chimpanzee and human clade
- premaxillary suture obliterated in adults
- lower molar protocristid grooves reduced
- axis of ear bones over 90°
- upper I2 similar in shape to I1
- shared substitutions at position 23 in the alpha haemoglobin chain and position 104 in the foetal globin A
- nuclear DNA: 1 insertion
  3 deletions
- mtDNA: 4 transversions
  7 transitions

Branch 7b Synapomorphies of the chimpanzee and gorilla (African ape) clade
- trigonid basin of lower molars enlarged
- six sacral vertebrae present
- 10 characters related to knuckle-walking complex
- 4 characters related to enamel structure complex
- ribosomal RNA: 1 deletion
- nuclear DNA: 1 transition
  1 deletion
- mtDNA: 8 transitions

condition which is therefore assumed to be primitive. The two events shared by chimpanzees and gorillas therefore appear to be synapomorphies; and similarly, the four deletions or insertions shared by chimpanzees and humans also appear to be synapomorphies. Clearly one of these sets of characters must be due to parallel change.

The African apes and humans are the only triplet of species sharing mutations. They share 11 nucleotide substitutions, of which three are transversions, and three deletions. For all of these, the alternative condition is found in the orang utan, Old World monkeys and New World monkeys, and this is assumed to be primitive for the anthropoids. The characters in the African apes and humans are therefore assumed to be synapomorphies, providing good evidence for this clade (to the exclusion of the orang utan). The great apes and humans are the only group of four species sharing mutations. They share a large number of substitutions, of which many are transversions, as well as deletions and insertions. (The exact numbers are not comparable with the figures for pairwise similarities for the great apes because for some of the sequence data published by Koop *et al.* (1986) the mammalian outgroup condition is unknown, and for other parts of the sequence the lemur outgroup is also missing). The alternative condition for these is seen in the Old World and New World monkeys, and this outgroup condition is assumed to be primitive for the anthropoids.

Applying these same principles to the sequence of 896 mtDNA nucleotides sequenced by Brown *et al.* (1982), Andrews (1986) found that chimpanzees and humans shared the greatest number of uniquely shared substitutions. A total number of 11 was counted, four of which on inspection proved to be transversions, and this compares with eight substitutions, all transitions, shared by chimpanzees and gorillas (and a further two shared by gorillas and humans). Thus although this provides some support for an African ape clade, the weight of the evidence must favour pairing of chimpanzees and humans on the basis of the four shared transversions. Twenty substitutions are shared by chimpanzees, humans and gorillas, including five transversions, and as these are not present in orang utans, gibbons and the mammalian outgroups, they are presumed to be shared derived characters supporting an African ape and human clade.

In summary, from those molecular data sets that we have been able to analyse cladistically certain conclusions emerge. Firstly, the great apes and humans share a number of derived features which are best interpreted as indicating that they form a clade. Within that clade the African apes and humans share a number of derived character states which are not present in *Pongo* (Figure 3) so that an African ape and human clade appears well substantiated. Relationships within the African ape and human clade are less clear: a chimpanzee/human clade is supported by 17 apparent synapomorphies, ten of which are more or less robust; an African ape clade is supported by 11 apparent shared derived characters, only two of which are of comparable robustness to the alternative grouping. If each genetic event is counted equally then a chimpanzee/human clade appears the most likely, although in view of the high levels of homoplasy present whichever branching alternative is adopted, this conclusion can hardly be said to be strongly indicated.

Cladistic analysis of morphological data

The morphological evidence has recently been reviewed by Andrews (1985) and Martin (1986). The details of our analysis are essentially as given in those papers and will not be repeated here. A selection of the characters which define the recognized hominoid clades, and which are most relevant to the available fossil evidence, is given in the caption to Figure 3. The hominoid clade is defined by 35 morphological characters which we interpret as synapomorphies, with the addition of molecular characters. The great ape and human clade is defined by a further 25 shared derived morphological characters and can additionally be substantiated through molecular data. Within this latter clade, the morphological evidence may be said to weakly indicate an African ape and human clade on the basis of the shared loss of an os centrale, and possession of a supraorbital torus and a frontal ethmoidal sinus, although the significance of the last character is dubious in the light of recent studies (Ward, personal communication). We follow Martin (1986) and Schwartz (1986) in interpreting the subnasal complex indicated by Ward & Pilbeam (1983) as a shared derived condition for the African ape and human clade as in fact being the likely ancestral pattern for the great ape and human clade. It would be fair to say that the morphological data do not strongly support an African ape and human clade at present. This will of course present us with considerable problems in recognizing the relationships of any fossil taxa which fall at or close to the separation of African apes, humans and orang utans.

The morphological data are much clearer as concerns the recognition of an African ape clade (and a human/orang clade). In particular, two character complexes, forelimb features associated with knuckle walking, and enamel formation and structure associated with enamel thickness on the molar teeth, seem to us to provide strong evidence for a period of common ancestry for chimpanzees and gorillas which did not include ancestors of either humans or orang utans. As we have decided to weight these two multiple character complexes rather heavily compared to, for example, the shared possession of a spatulate I^2 by chimpanzees and humans (Martin, 1986) we feel that our rationale requires a brief explanation.

In two previous analyses, a large number of shared derived characters were identified between chimpanzees and gorillas all of which related to the forelimb, and particularly the hand and wrist (Martin, 1986; Andrews, 1987). These can be clearly shown to be part of a character complex which is functionally related to knuckle walking in these taxa (Tuttle, 1967), and consequently we feel that they should be treated as a single character. Since there are a large number of components to this complex, and since they are all identical in chimpanzees and gorillas, it is our belief that they provide strong support for the common ancestry of the two African apes. We have explained above, and more details may be found in Andrews (1987), that although the characters are functionally correlated, the chances of the entire complex being evolved in parallel are very small when all (numerous) elements are identical. It follows that we consider it unlikely that the knuckle-walking complex of characters was present in the common ancestor of African apes and humans as some have implied. Moreover there is a second character complex from an entirely unrelated functional area which also supports this interpretation.

Recent studies of enamel thickness in hominoid primates have shown that thickness is a character which is not only correlated but causally related, in a developmental sense, to both rate of secretion of the enamel, and the resultant enamel prism packing pattern (Martin, 1983, 1985). As such, it is perhaps best to treat these apparently separate characters also as part of a character complex rather than to treat them separately as was done by Martin (1986). This is particularly important in the light of analyses of the same data set by other workers (e.g., Schwartz, 1986) who have developed schemes which are internally inconsistent in terms of the known consequences for enamel thickness of developmental parameters such as secretion rate and enamel prism packing patterns. The enamel complex data supports the identification of a great ape and human clade defined by thick fast-formed (Pattern 3 enamel of Boyde, 1967). Within that clade it also defines an African ape clade defined by secondarily reduced, slow formed (Pattern 1) thin enamel in chimps and gorillas. The data do not support the presence of an African ape and human clade, nor have they ever been used to indicate such (cf. Schwartz, 1986) and nor do they support an orang utan/human clade. As the development of enamel is very well understood in hominoid primates we feel quite confident that the shared developmental pattern and resultant thin enamel shared by chimps and gorillas is evidence for a shared period of common ancestry.

When the molecular and morphological evidence are combined in a cladistic analysis the cladogram shown in Figure 3 appears to us to be the most parismonious interpretation. It requires some homoplasy, particularly in molecular characters but perhaps we should not be surprised to find this. The cladogram involves relatively little homoplasy for the morphological evidence and perhaps the only real surprises are that the clear shared derived similarities at the molecular level among African apes and humans are not strongly reflected in the morphological analysis and that the morphological similarities between African apes are not reflected at the molecular level. Hominoidea is well defined by morphology and molecular characters, similarly the great ape and human clade. The African ape and human clade is particularly well defined on molecular evidence but finds some support in morphological characters as well. The African ape clade is defined largely on the basis of two character complexes which have been weighted in this analysis as explained above.

Fossil evidence

Up to this point we have restricted our attention to the relationships of the living hominoids. Arguments for and against various interpretations of branching points in hominoid phylogeny have been given, and we will now consider the placement of fossils in relation to these branching points. In this context, the fossils are seen as providing corroboration of the branching sequence already established, adding additional characters either as potential synapomorphies or for filling in detail on the ancestral states at various nodes, and, most importantly, providing a time scale in order to derive a phylogeny out of the sets of relationship. Figures 1 and 3 have been shown to illustrate hominoid relationships, and by adding fossils to Figure 3 the phylogeny in Figure 4 has been obtained. Each of the branching points in Figure 4 will now be considered, with a justification for the inclusion of the fossil taxa at these points and brief comments made about the significance of these relationships to hominoid phylogeny.

Branch 1 on Figure 4 gives the position for any stem hominoids that may be recognized: that is, fossil hominoids which branched off after the division between living cercopithecoids and hominoids but which cannot be placed into any one of the hominoid clades. In Figure 4 the early Miocene genus *Proconsul* is identified as a stem hominoid, following Andrews (1985) and Martin (1986) but in contrast to Harrison (1987). The species of this genus are well represented from several sites in East Africa (Andrews, 1978). Their skulls have enlarged maxillary sinuses, reduced heteromorphy of upper premolars, and reduced lower P3 height, while their postcranial skeleton has a scapula with elongated vertebral border, rounded humeral head, and probably reduction in lumbar and increase

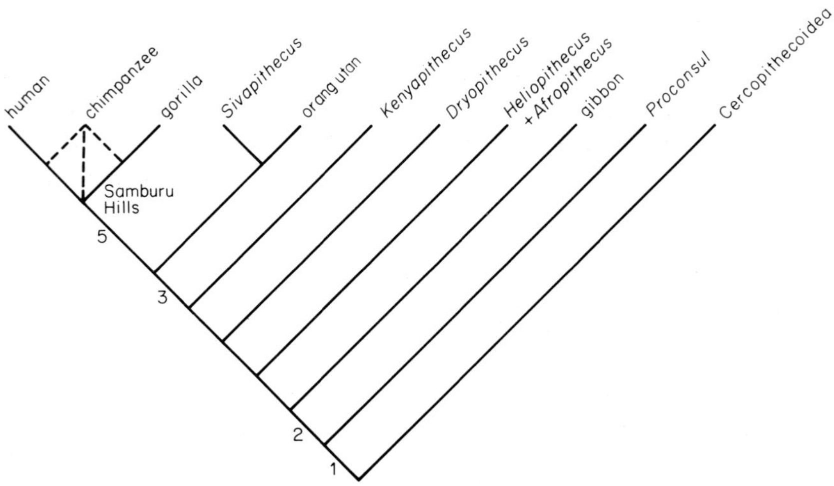

Figure 4. Cladogram showing the relationships discussed in the text for the extant hominoids and fossil taxa whose positions have been assessed.

in sacral vertebrae. These are the characters that have been shown to define Hominoidea on branch 2 of Figure 1, but none of the characters listed in Figure 1 for branches 3 (great ape and human clade) or 4 (gibbon clade) are present in *Proconsul*. It may be concluded, therefore, that *Proconsul* can best be regarded as a stem hominoid and that its presence in Africa at least 22 million years ago (Andrews *et al*., 1981) gives a minimum time for hominoid origins.

Branch 2 of Figure 4 is defined exactly as for branch 2 of Figure 1, and this includes a number of characters which define the living species of Hominoidea and some fossils, but which are not present in *Proconsul*. Most of these have not been listed in the present paper, but may be found in detail in Andrews (1985, table 2). Similarly, branch 3 of Figure 4 is defined as for branch 3 of Figures 1 and 3, and this includes a selection of the defining characters of the great ape and human clade which are given in full in table 2 of Martin (1986). The intervening branches between these encompass the stem group of great apes and humans, and four fossil hominoid genera have been identified as part of this group. These will now be discussed in turn.

There is some uncertainty about the relative positions on Figure 4 of some of these stem group genera. *Afropithecus* and *Heliopithecus* are two genera that share a number of plesiomorphies with the great ape and human clade and may be considered on present evidence to be closely related, although this requires confirmation from more complete evidence, especially for the latter genus. The former genus comes from Kalodirr in northern Kenya and is reliably dated to between 17 and 18 Ma (Leakey & Leakey, 1986) and includes material from Buluk, Kenya, while the latter comes from Ad Dabtiyah in Saudi Arabia and has a faunal age of around 17 Ma (Andrews & Martin, 1987). Saudi Arabia forms part of the African plate, and in the early Miocene it must be considered as palaeogeographically part of Africa. Material from Maroto, Uganda may have affinities with *Afropithecus*, *Heliopithecus* and *Kenyapithecus* (Martin, 1981; Andrews & Martin, 1987). In the skull of *Afropithecus*, the orbits are higher than broad, the premaxilla is lengthened but appears to lack the stepped floor of the nasal chamber brought about by the superimposition of the premaxilla on the maxilla considered here to be characteristic of the great apes and humans. The lower P3 is robust (not known for *Heliopithecus*) and the upper premolars are greatly enlarged; the canines are robust and evidently had reduced honing; the upper molars have reduced cingula, although traces remain, and at least in *Heliopithecus* there is evidence for increase in enamel thickness: the molar enamel is of an intermediate thickness predicted by Martin (1985) but not actually identified in any hominoid before its discovery in this fossil genus. The mandible in *Afropithecus* is of great depth and robustness, and the symphysis in particular is very deep and the massive inferior torus is low down on the symphysis. For most or all of these characters, the two fossil genera share the same condition with the great ape and human clade as defined in Figure 1.

These two fossil genera identify a minimum time for the divergence of the gibbons from the great ape and human clade. No gibbons are known from the fossil record, so that it is not possible to say anything about this clade at present (Fleagle, 1984), but the presence in *Heliopithecus* and *Afropithecus* of several great ape and human synapomorphies for which gibbons are primitive makes it possible to conclude that the divergence of the two clades must already have occured prior to the 17 to 18 Ma time period during which the two fossil apes lived. During the early Miocene and the early part of the middle Miocene, fossil hominoids have been found only in Africa, so that as well as the whole group originating in Africa it seems likely that the divergence between the great apes and humans on the one hand and gibbons on the other also took place in Africa.

Another genus from Africa, *Kenyapithecus*, also appears to share a number of great ape and human synapomorphies. This genus, originally named by Louis Leakey (1962), was synonymized with "*Ramapithecus*" by Simons & Pilbeam (1965), but is now increasingly recognized as a distinct genus. We have used data from Fort Ternan, Majiwa, Kaloma and Maboko Island, Kenya in assessing its affinities. It is known only from fragmentary remains recovered from middle Miocene deposits in Kenya, but it shares many of the great ape and human characters listed above and in Figure 1 branch 3, particularly the enlarged and robust premolars, reduction in upper premolar heteromorphy exceeding that seen in *Heliopithecus*, extremely robust mandible, and molars with thick enamel. In the other characters listed, *Kenyapithecus* either shares the great ape and human characters with morphology similar to the condition in the other fossil genera, or the characters are unknown because it is relatively less complete. In no case does it retain an ancestral hominoid condition for which *Heliopithecus* or *Afropithecus* are derived, while on the contrary it shares great ape and human character states for which the other genera are either primitive or less advanced. It seems reasonable to conclude, therefore, that it is more closely related to that clade than are the other fossils, and so it is shown on Figure 4.

So far the evidence is unambiguous, but consideration of the fossil genus, *Dryopithecus*, raises some problems. There is first of all the problem of what is meant by the name: we include the type material from France and the remains from Germany, Austria and Spain referred to the genus by Simons & Pilbeam (1965), and we also include the specimens from Hungary assigned by Kretzoi (1975) to two other genera (Martin & Andrews, 1982) except for one large mandible. These are all late middle Miocene to late Miocene in age, and on this definition also the genus is exclusively European. In comparison with the other stem group genera just described, therefore, *Dryopithecus* differs both in terms of geography, being non-African, and time, being several million years later in time.

In terms of morphology, *Dryopithecus* shares a number of character states with the other fossil genera and with the great ape and human clade. For some characters, however, it is less advanced towards the condition described for branch 3 of Figure 1, while in others it is more advanced. Equivalence is seen in the relatively high narrow orbits, reduction in bilateral compression of the lower P3. Characters in which it is less advanced than the other fossil genera include the lack of alveolar prognathism and short premaxilla, the relatively small upper premolars, the gracile canines and the relatively gracile mandibular corpus. In terms of enamel thickness little is known for *Dryopithecus* except that it shows the pattern of dentine separation wear (Martin, 1983) which has been shown to characterize species which do not have thick enamel. It is intermediate between *Heliopithecus* and *Kenyapithecus* (i.e. more derived than the former but less "advanced" than the latter) in the reduction in upper premolar heteromorphy and, probably, enamel thickness. Characters in which it is more derived than all of the other genera are the enlargement of the maxillary sinuses, the stepped floor of the nasal chamber with reduced incisive fossa (intermediate condition between the ancestral hominoid condition seen in *Proconsul* and the great ape and human condition), and the reduction in upper molar cingula. These contradictions in the morphological evidence make it difficult to place *Dryopithecus*, but we have put it in an intermediate position because of its lack of thick molar enamel (seen in *Kenyapithecus*) and subnasal morphology: see Figure 4.

It has already been observed that there is no fossil evidence presently available for branch 4 leading to the gibbon clade on Figure 3. Similarly, there is an absence of fossil evidence on the African ape and human clade (branch 5 Figure 3). No fossils that can unequivocally be placed within this clade are known before the time of *Australopithecus afarensis*, and none at all are known for the African apes. One possibility that can be considered briefly is the maxilla from Samburu Hills (Ishida *et al.*, 1984), which has some similarities with gorillas in details of its dentition. The premolars are very elongated, which has been shown to be a great ape and human character, but the degree of elongation is greater than in any of the fossil hominoids just considered and is identical to the present great ape condition, so that it appears to constitute an advance with respect to the other fossils. The molars are relatively elongated also, to a degree seen only in gorillas among the great apes and humans, and the occlusal ridges of the molars are poorly defined, with the cusps separated by deep invaginations, again as seen only in gorillas. The significance of these similarities to gorillas is uncertain: they appear to be uniquely shared character states, but the possibility of their being primitive cannot be excluded on present evidence. The enamel also appears to be thick on the Samburu Hills specimen, which would be predicted on our model (Figure 1), but which makes it difficult to predict homology in the other characters of the teeth. For the moment, therefore, and until more fossils of this type are available for study, we stop short of linking it with the gorilla or even with the African ape clade, although there is a strong probability that this is where it belongs.

One last fossil group will be considered here, in relation to branch 6 of figure 3. This is the orang utan clade, and it has been well demonstrated that the fossils from Indo-Pakistan and Turkey assigned to the genus *Sivapithecus* (including "*Ramapithecus*") belong to this clade (Andrews & Cronin, 1982; Ward & Pilbeam, 1983; Schwartz, 1984). The orang utan has a number of character states that are unique among catarrhine primates, and these can be identified as autapomorphies with some degree of confidence. Their presence in *Sivapithecus* from Indo-Pakistan and from Sinap, Turkey must therefore be considered strong evidence for their relationship with the orang utan. It has been further claimed by Schwartz (1984) that this whole clade is the sister group to humans, which has the merit of accommodating the previously described similarities between humans and "*Ramapithecus*"

(Simons & Pilbeam, 1965), but our interpretation of the possible link between mandibular robusticity and enamel thickness (Martin & Andrews, 1984), the two most striking similarities between humans and *"Ramapithecus"*, make this accommodation unnecessary. In view of the strong support for closer relationship between humans and African apes, with the orang utan as the outgroup to these, from amino acid and DNA sequencing data (see above), we do not presently consider the hypothesis of relationship between humans and orang utans (including *Sivapithecus*) to be valid.

Acknowledgements

We are grateful to Fred Grine for inviting us to present this work at the Primate Phylogeny symposium at the VI Hennig Society meeting. We are also grateful to David Daegling, Eric Delson, John Fleagle, Fred Grine, Masami Hasegawa, Jon Marks and an anonymous referee for comments on the manuscript for this paper and to Chris Stringer and Jeff Schwartz for much entertaining discussion and disagreement. Much of the data analysed here come from a continuing collaboration among PA, Colin Groves and Jeff Schwartz, and we are grateful for their permission to use the joint data here.

References

Andrews, P. (1978). A Revision of the Miocene Hominoidea of East Africa. *Bull. Br. Mus. nat. Hist. (Geol.)* **30,** 85–224.

Andrews, P. (1985). Family group systematics and evolution among catarrhine primates. In (E. Delson, Ed.) *Ancestors: the Hard Evidence* pp. 14–22. New York: Alan R. Liss.

Andrews, P. (1986). The molecular evidence for catarrhine evolution. In (B. Wood, L. Martin & P. Andrews, Eds) *Major Topics in Primate and Human Evolution*, pp. 107–129. Cambridge: Cambridge University Press.

Andrews, P. (1987). Aspects of hominoid phylogeny. In (C. Patterson, Ed.) *Molecules and Morphology in Evolution—Conflict or Compromise*, pp. 23–53. Cambridge: Cambridge University Press.

Andrews, P. & Cronin, J. (1982). The relationships of *Sivapithecus* and *Ramapithecus* and the evolution of the orang utan. *Nature* **297,** 541–546.

Andrews, P. & Groves, C. P. (1975). Gibbons and brachiation. In (D. M. Rumbaugh, Ed.) *Gibbon and Siamang*, Vol. 4, pp. 167–218. Basel: Karger.

Andrews, P., Harrison, T., Martin, L. & Pickford, M. (1981). Hominoid primates from a new Miocene locality named Meswa Bridge in Kenya. *J. hum. Evol.* **10,** 123–128.

Andrews, P. & Martin, L. (1987). The phyletic position of the Ad Dabtiyah hominoid. *Bull. Br. Mus. nat. Hist. (Geol.)* **41,** 383–393.

Boyde, A. (1967). The development of enamel structure. *Proc. r. Soc. Med.* **60,** 13–18.

Bishop, M. J. & Friday, A. E. (1986). Molecular sequences and hominoid phylogeny. In (B. Wood, L. Martin & P. Andrews, eds) *Major Topics in Primate and Human Evolution*, pp. 150–156. Cambridge: Cambridge University Press.

Brown, W. M., Prager, E. M., Wang, A. & Wilson, A. C. (1982). Mitochondrial DNA sequences of primates: tempo and mode of evolution. *J. molec Evol.* **18,** 225–239.

Delson, E. & Andrews, P. (1975). Evolution and interrelationships of the catarrhine primates. In (W. P. Luckett & F. S. Szalay, Eds) *Phylogeny of the Primates*, pp. 405–446. New York: Plenum Press.

Doolittle, R. F., Wooding, G. L., Lin, Y. & Riley, M. (1971). Hominoid evolution as judged by fibrinopeptide structures. *J. molec. Evol.* **1,** 74–83.

Ferris, S. D., Wilson, A. C. & Brown, W. M. (1981). Evolutionary tree for apes and humans based on cleavage maps of mtDNA. *Proc. natl Acad. Sci.* **78,** 2432–2436.

Fleagle, J. G. (1984). Are there any fossil gibbons? In (H. Preuschoft, D. J. Chivers, W. Y. Brockelman & N. Creel, Eds) *The Lesser Apes*, pp. 431–447. Edinburgh: Edinburgh University Press.

Goldman, D., Rathna Giri, P., & O'Brien, S. J. (1987). A molecular phylogeny of the hominoid primates as indicated by two-dimensional protein electrophoresis. *Proc. natl Acad. Sci.* **84,** 3307–3311.

Goodman, M. (1963). Man's place in the phylogeny of the primates as reflected in serum proteins. In (S. L. Washburn, Ed.) *Classification and Human Evolution*, pp. 204–234. Chicago: Aldine.

Goodman, M. (1976). Towards a genealogical description of the primates. In (M. Goodman & R. E. Tashian, Eds) *Molecular Anthropology*, pp. 321–353. New York: Plenum Press.

Goodman, M., Braunitzer, G., Stangl, A. & Schrank, B. (1983). Evidence on human origins from haemoglobins of African apes. *Nature* **303,** 546–548.
Groves, C. P. (1986). Systematics of the Great Apes. In (D. R. Swindler & J. Erwin, Eds) *Comparative Primate Biology*, Vol. 1, *Systematics, Evolution and Anatomy*, pp. 187–217. New York: Alan R. Liss.
Harrison, T. (1982). Small-bodied apes from the Miocene of East Africa. Ph.D. thesis. University of London.
Harrison, T. (1987). The phylogenetic relationships of the early catarrhine primates: A review of the current evidence. *J. hum. Evol.* **16,** 41–80.
Hasegawa, M., Yano, T. & Kishino, H. (1984). A new molecular clock of mtDNA and the evolution of hominoids. *Proc. Japan Acad. Sci.* **60,** 95–98.
Hixson, J. E. & Brown, W. M. (1986). A comparison of the small ribosomal RNA genes from the mtDNA of the great apes and humans: sequence, structure, evolution and phylogenetic implications. *Mol. Biol. Evol.* **3,** 1–18.
Ishida, H. Pickford, M., Nakaya, H. & Nakano, Y. (1984). Fossil anthropoids from Nachola and Samburu Hills, Samburu District, Kenya. *Afr. Stud. Monog., Kyoto* **2,** 73–85.
Kluge, A. G. (1983). Cladistics and the classification of the great apes. In (R. L. Ciochon & R. S. Corruccini, Eds) *New interpretations of ape and human ancestry*, pp. 151–177. New York: Plenum Press.
Koop, B. F., Goodman, M., Peilin Xu, Chan, K. & Slightom, J. L. (1986). Primate n-globin DNA sequences and man's place among the great apes. *Nature* **319,** 234–238.
Kretzoi, M. (1975). New ramapithecines and *Pliopithecus* from the lower Pliocene of Rudabanya in north-eastern Hungary. *Nature* **257,** 578–581.
Leakey, L. S. B. (1962). A new lower Pliocene fossil primate from Kenya. *Ann. Mag. nat. Hist.* **4,** 689–696.
Leakey, R. E. & Leakey, M. G. (1986). A new Miocene hominoid from Kenya. *Nature* **324,** 143–145.
Marks, J. (1983). Hominoid cytogenetics and evolution. *Yb. phys. Anthrop.* **26,** 131–159.
Martin, L. (1981). New specimens of *Proconsul* from Koru, Kenya. *J. hum. Evol.* **10,** 139–150.
Martin, L. (1983). The relationships of the later Miocene Hominoidea. PhD thesis, University of London.
Martin, L. (1985). Significance of enamel thickness in hominoid evolution. *Nature* **314,** 260–263.
Martin, L. (1986). Relationships among extant and extinct great apes and humans. In (B. Wood, L. Martin & P. Andrews, Eds) *Major Topics in Primate and Human Evolution*, pp. 161–187. Cambridge: Cambridge University Press.
Martin, L. & Andrews, P. (1982). New ideas on the relationships of the Miocene hominoids. *Primate Eye* **18,** 4–7.
Martin, L. & Andrews, P. (1984). The phyletic position of *Graecopithecus freybergi* Koenigswald. *Cour. Forsch. Inst. Senckenberg* **69,** 25–40.
Romero-Herrera, A. E., Lehman, H. Castillo, O., Joysey, K. A. & Friday, A. E. (1976). Myoglobin of the orang utan as a phylogenetic enigma. *Nature* **261,** 162–164.
Romero-Herrera, A. E., Lehman, H., Joysey, K. A. & Friday, A. E. (1978). On the evolution of myoglobin. *Phil. Trans. r. Soc.* **283,** 61–163.
Sarich, V. M. & Wilson, A. C. (1967). Immunological time scale for hominid evolution. *Science* **158,** 1200–1203.
Sarich, V. M. & Cronin, J. E. (1976). Molecular systematics of the primates. In (M. Goodman & R. E. Tashian, Eds) *Molecular Anthropology*, pp. 140–170. New York: Plenum Press.
Schwartz, J. H. (1984). On the evolutionary relationships of humans and orang utans. *Nature* **308,** 501–505.
Schwartz, J. H. (1986). *The Red Ape*. Boston: Houghton Mifflin Co.
Scott, A. F., Heath, P., Trusko, S., Boyer, S. H., Prass, W., Goodman, M., Czelusniak, J., Chang, L.-Y. E. & Slightom, J. L. (1984). The sequence of the gorilla foetal globin genes: evidence of multiple gene conversions in human evolution. *Mol. Biol. Evol.* **1,** 371–389.
Sibley, C. G. & Ahlquist, J. E. (1984). The phylogeny of the hominoid primates, as indicated by DNA–DNA hybridization. *J. molec. Evol.* **20,** 2–15.
Simons, E. L. & Pilbeam, D. R. (1965). Preliminary revision of the Dryopithecinae (Pongidae, Anthropoidea). *Folia primat.* **3,** 81–152.
Tashian, R. E., Goodman, M., Ferrell, R. E. & Tanis, R. J. (1976). Evolution of carbonic anhydrase in primates and other mammals. In (M. Goodman & R. E. Tashian, Eds) *Molecular anthropology*, pp. 301–319. New York: Plenum Press.
Templeton, A. R. (1983). Phylogenetic inference from restriction endonuclease cleavage site maps with particular reference to the evolution of humans and apes. *Evolution* **37,** 221–244.
Tuttle, R. H. (1967). Knuckle-walking and the evolution of hominoid hands. *Am. J. phys. Anthrop.* **26,** 171–206.
Ward, S. C. & Pilbeam, D. R. (1983). Maxillofacial morphology of Miocene hominoids from Africa and Indo-Pakistan. In (R. L. Ciochon & R. S. Corruccini, Eds) *New Interpretations of Ape and Human Ancestry*, pp. 211–238. New York: Plenum Press.
Yunis, J. J. & Prakash, D. (1982). The origin of man: a chromosomal pictorial legacy. *Science* **215,** 1525–1530.

A. T. Chamberlain
B. A. Wood

Hominid Palaeontology Research Group, Department of Human Anatomy and Cell Biology, The University of Liverpool, P.O. Box 147, Liverpool, L69 3BX, England

Received 20 January 1987
Revision received 26 June 1987
and accepted 2 July 1987

Publication date November 1987

Keywords: hominids, *Australopithecus, Homo,* phylogeny, cladistics.

Early hominid phylogeny

This paper describes a quantitative cladistic study of hominid phylogeny. Measurements between homologous landmarks were taken from a variety of fossil hominids and from 140 modern primate specimens. The measurements were confined to the cranium, mandible and dentition. The effect of size was removed from the data using an adaptation of Kazmierczack's transformation, and the transformed data were coded directly into discrete, ordered character states. Cladograms were constructed for different OTU combinations, using the PAUP phylogeny program to find minimum length (maximum parsimony) trees. The trees were rooted by including a single hypothetical outgroup which was calculated from a previous, separate, analysis of six extant primate taxa.

The most parsimonious cladogram for conventionally-defined hominid taxa is consistent with the phylogenetic hypothesis advanced by Kimbel *et al.* (1984). Subdivision of early *Homo*, according to criteria that emphasize the size variation in this group, has little impact on the phylogenetic relationships of the group. An alternative division of early *Homo*, based on a phenetic analysis of shape, suggests that one subgroup is primitive while the other has affinities with the "robust" australopithecines.

Journal of Human Evolution (1987) **16,** 119–133

Introduction

Despite the wealth of fossil hominid material, and the intensity with which such material has been studied, hominid phylogeny is still far from being resolved. Major phylogenetic problems remain to be answered and these concern not only the relationships between individual species, but also the ordering of the major branching events that resulted in the diverse range of taxa currently recognized by those who study the fossil evidence. The uncertainty that prevails in hominid systematics is reflected both in the proliferation of alternative phylogenetic schemes (cf. Johanson & White, 1979; Tobias, 1980; Skelton *et al.*, 1986), and in their lack of success in predicting the limits of existing taxa or the nature of ancestral forms (Delson, 1986; Walker *et al.*, 1986). In part, this state of affairs can be attributed to the uneven temporal distribution of the evidence as well as to the lack of consensus over the taxonomic affinities of the specimens that are presently known. We feel, however, that hominid phylogeny will continue to be reactive to, rather than predictive of future fossil discoveries until sufficient attention is paid to methods of phylogenetic reconstruction and unless these methods are applied objectively and rigorously to the available fossil evidence.

The "explosion" in the rate of recovery of fossil hominids has coincided with a vigorous debate about methods of phylogenetic reconstruction (Charig, 1982; Hull, 1984). Several authors have discussed hominid systematics from a cladistic perspective (Eldredge & Tattersall, 1975; Bonde, 1977; Delson *et al.*, 1977; Tattersall & Eldredge, 1977; Johanson & White, 1979; White *et al.*, 1981; Skelton *et al.*, 1986), but there have been few cladistic studies carried out on original data (Corruccini & McHenry, 1980; Kimbel *et al.*, 1984; Wood & Chamberlain, 1986). Since the latter study was undertaken we have accumulated additional measurements of fossil hominids, and have acquired identically-defined measurements for a range of extant primates. We have also adopted a novel method of generating character states from linear measurement data. Furthermore, powerful and

0047–2484/87/010119 + 15 $03.00/0

sophisticated phylogenetic programs are now widely available for microcomputers (Fink, 1986). In this contribution we report the results of a new cladistic analysis of hominids that takes advantage of the above developments.

Materials and methods

Ingroup (hominids)

The fossil hominid specimens included are listed in Table 1. They comprise the more complete cranial and mandibular remains of a range of Plio-Pleistocene hominids: no postcranial remains were included. The specimens were initially assigned to

Table 1 **Fossil hominid specimens included in the analysis, assigned to conventional taxa**

A. afarensis	Hadar:	AL 128-23, 198-1, 266-1, 277-1, 288-1i, AL 333w32/60, 400-1a, 200-1a, 333-1, 333-45
	Laetoli:	LH 4
A. africanus	Sterkfontein:	Sts 5, 7, 36, 52, 71
	Makapansgat:	MLD 18, 40
A. robustus	Swartkrans:	SK 12, 23, 34, 46, 48, 52
	Kromdraai:	TM 1517
A. boisei	Koobi Fora:	KNM-ER 406, 732, 729, 818, 3230
	Olduvai:	OH 5
	Peninj:	Peninj 1
H. habilis	Koobi Fora:	KNM-ER 1470, 1802, 1805, 1813, 3734
	Olduvai:	OH 7, 13, 24
	Swartkrans:	SK 847
	Sterkfontein:	Stw 53
H. erectus	China:	Zhoukoudian H1, K1 (mandibles)
	Indonesia:	Sangiran 1, 4, 9
	Africa:	KNM-ER 730, 992, 3733, 3883, OH 22, BK 67, BK 8518, SK 15, Ndutu

Note: both *H. habilis* and *H. erectus* are defined *sensu lato* (cf. Howell, 1978).

Table 2 **Specimens of early *Homo* common to both Stringer's study (Stringer, 1986) and to the present study**

Specimen	Stringer 1986 classification	Present study classification
OH 7	Group 1	*H. habilis*
OH 13	Group 2	*H. habilis*
OH 24	Group 1	*H. habilis*
KNM-ER 1470	Group 1	*H.* sp.
KNM-ER 1802	Group 1	—
KNM-ER 1805	Group 2	*H.* sp.
KNM-ER 1813	Group 2	*H.* sp.
KNM-ER 992	Group 2	—

Note: two mandibles (KNM-ER 992 & 1802) that have been attributed by some authors to *H. habilis* were not referred to *H. habilis* or to *Homo* sp. in the present study. According to our metrical data, neither of these Koobi Fora isolated mandibles resemble *H. habilis*; their closest affinities are with *Homo* sp. (as represented by KNM-ER 1805) and with *H. erectus* (as represented by KNM-ER 730) (Chamberlain, in preparation).

conventionally-defined taxa (Table 1: cf. Howell, 1978; Day, 1986). An alternative classification of fossil material conventionally attributed to *H. habilis* has been proposed by Stringer (1986). Specimens considered by Stringer, and also included in the present study, are listed in Table 2. The same table also gives the classification of those specimens following a phenetic analysis of fossil hominids undertaken by one of us (Chamberlain, in preparation). In addition to the fossil hominid specimens, a balanced-sex group of 20 adult modern *Homo sapiens* crania was randomly sampled from the extensive collection of African crania in the Department of Anatomy, University of the Witwatersrand.

Outgroup (primates)

Outgroup taxa are groups related to, but not included within, the monophyly containing the taxa under investigation. Higher primates constitute the most relevant outgroup taxa for the purposes of the present study. The following species were selected from among the subfamilies of the cercopithecoid and hominoid primates: *Papio anubis, Colobus badius, Hylobates lar, Pongo pygmaeus, Gorilla gorilla* and *Pan troglodytes*. The catarrhine primates have been shown by molecular comparisons to be most closely related to modern humans, and in addition their morphology is such that almost all of the measurements employed in the present study are homologous between different animals. The use of more distantly-related primates, such as ceboid monkeys and prosimians, was precluded because some of the measurements used in this study (e.g., length of premolar tooth row, distance between tympanic plates) would have been either unavailable or not homologous in such a selection of primates. For each outgroup species, 10 male and 10 female adult crania were sampled.

Sampling of characters

The selected measurements (Table 3) were chosen in order to render in quantitative form a range of morphological characters known to be variable among hominids. Measurements were limited to the cranium and mandible: isolated postcranial elements are common in the hominid fossil record, but in many instances their referral to species cannot be agreed upon (Kennedy, 1983; Trinkaus, 1983; Johanson *et al.*, 1987). The skull was divided into five functional/anatomical regions to ensure a degree of uniformity in the distribution of characters measured (cf. Wood & Chamberlain, 1986). The dental measurements were duplicated on upper and lower dentitions. Although this might appear to introduce some redundancy into the data set, upper and lower dental data were rarely available for the same fossil individual and morphological correlations between some upper and lower antimeres are complex and variable among primate taxa.

Most landmarks and measurement definitions made use of conventional morphometric definitions (see Table 3); detailed definitions are given in Chamberlain (in preparation). A few conventional landmarks and measurements were re-defined in the present study in order to allow easier identifications on fossils, or to allow for morphological variation across the range of primate species sampled. All measurements were made by one observer (ATC). Measurements available bilaterally on a specimen were averaged. One complete specimen of *Pan troglodytes* was remeasured on a separate occasion to provide an estimate of intra-observer error. The mean difference between the original and replicated measurements on this specimen amounted to 2·3% of the average measurement value; the median difference was 1·1%.

Table 3 **List of measurements**

No.	Symbol	Description
1	B1	Intertympanic breadth (TP-TP)
2	B2	Breadth between carotid canals (CC-CC)
3	B3	Breadth between petrosal apices (PA-PA)
4	B4	Breadth between foramina ovale (FO-FO)
5	B5	Breadth between infratemporal crests (ITS-ITS)
6	B6	Mandibular fossa breadth (entoglenoid-lateral margin)
7	B7	Tympanic length (CC-TP)
8	B8	Petrosal length (CC-PA)
9	B9	Common carotid to foramen ovale (CC-FO)
10	B10	Common carotid to infratemporal crest (CC-ITS)
11	B11	Foramen magnum length
12	B12	Foramen magnum breadth
13	B13	Temporal fossa length
14	B14	Temporal fossa breadth
15	B15	Basal length (opisthion subtense from ITS/ITS)
16	B16	Basioccipital length (basion-sphenoccipital synchrondrosis)
17	V1	Parietal thickness at lambda
18	V2	Minimum frontal breadth
19	V3	Maximum biparietal breadth
20	V4	Biporionic breadth
21	V5	Cranial length (glabella-opisthocranion)
22	V6	Frontal sagittal chord (nasion-bregma)
23	V7	Parietal sagittal chord (bregma-lambda)
24	V8	Parietal coronal chord (bregma-pterion)
25	V9	Occipital sagittal chord (lambda-opisthion)
26	V10	Frontal sagittal arc (nasion-bregma)
27	V11	Occipital sagittal arc (lambda-opisthion)
28	V12	Auricular height (vertex subtense from porion/porion)
29	V13	Posterior length (opisthocranion subtense from porion/porion)
30	F1	Upper facial breadth (between frontomalare temporale)
31	F2	Mid-facial breadth (zygion-zygion)
32	F3	Lower facial breadth (zygomaxillare-zygomaxillare)
33	F4	Interforaminal breadth (between infraorbital foramina)
34	F5	Biorbital breadth (ectoconchion-ectoconchion)
35	F6	Orbital breadth (dacryon-ectoconchion)
36	F7	Interorbital breadth (dacryon-dacryon)
37	F8	Maximum (lower) nasal bone breadth
38	F9	Maximum nasal aperture breadth
39	F10	Facial height (nasion-alveolare)
40	F11	Foraminal height (infraorbital foramen-P^4 alveolus)
41	F12	Lower orbital height (zygoorbitale-P^4 alveolus)
42	F13	Upper malar height (frontomalare temporale-M^2/M^3 alveolus
43	F14	Lower malar height (zygomaxillare-M^2/M^3 alveolus)
44	F15	Upper facial prognathism (porion-glabella)
45	F16	Lower facial prognathism (porion-alveolare)
46	F17	Malar prognathism (porion-zygomaxillare)
47	F18	Nasion subtense from frontomalare/frontomalare
48	F19	Subspinale subtense from zygomaxillare/zygomaxillare
49	M1	Mandibular corpus depth at symphysis
50	M2	Mandibular corpus depth at M_1
51	M3	Mandibular corpus depth at M_3
52	M4	Height above base of foramen spinosum
53	M5	Height above base of mental foramen
54	M6	Mandibular corpus breadth at symphysis
55	M7	Mandibular corpus breadth at M_1
56	M8	Anterior intercorpus breadth (C_1-C_1)
57	M9	Posterior intercorpus breadth (M_2-M_2)

Table 3—*continued*

No.	Symbol	Description
58	M10	Incisor alveolar length
59	M11	Premolar alveolar length
60	M12	Molar alveolar length
61	M13	Mesiodistal I_1
62	M14	Buccolingual I_1
63	M15	Mesiodistal C_1
64	M16	Buccolingual C_1
65	M17	Mesiodistal P_3
66	M18	Buccolingual P_3
67	M19	Mesiodistal M_1
68	M20	Buccolingual M_1
69	M21	Mesiodistal M_3
70	M22	Buccolingual M_3
71	P1	Maxilloalveolar length
72	P2	Maxilloalveolar breadth
73	P3	Anterior palatal breadth (C^1-C^1)
74	P4	Posterior palatal breadth (M^2-M^2)
75	P5	Palate depth at incisive fossa
76	P6	Palate depth at M^2/M^2
77	P7	Alveolare subtense from C^1/C^1
78	P8	Incisor alveolar length
79	P9	Premolar alveolar length
80	P10	Molar alveolar length
81	P11	Mesiodistal I^1
82	P12	Buccolingual I^1
83	P13	Mesiodistal C^1
84	P14	Buccolingual C^1
85	P15	Mesiodistal P^3
86	P16	Buccolingual P^3
87	P17	Mesiodistal M^1
88	P18	Buccolingual M^1
89	P19	Mesiodistal M^3
90	P20	Buccolingual M^3

Size standardization

Preliminary results obtained by comparing the raw data from extant primates indicated that unless the raw measurements were corrected for size differences between individuals and between taxa, no sensible taxonomic or phylogenetic affinities could be extracted from the data. In the absence of any correction for size we found that sexual dimorphism was the dominant influence on any classification. In our previous analysis (Wood & Chamberlain, 1986) we followed other authors (e.g., Kimbel *et al.*, 1984) in using ratios to control for the effects of size. Ratios are usually calculated by dividing one measurement by another measurement taken on the same individual, and their statistical properties are therefore influenced by the nature of the relationship between the denominator and numerator variables (Atchley *et al.*, 1976; Hills, 1978).

We avoid these potential difficulties by replacing the ratio denominator by a constant. The latter is also a measurement, defined in exactly the same terms as the numerator variable, but is is taken on a "standard" form. Since the resulting ratio no longer corrects for size differences we divide it by an additional term which expresses the average size ratio between the individual and the "standard" form. Thus the transformation for measurement a on individual i is given by Equation (1):

$$R_{a_i} = (a_i/a_s)/(X_i/X_s) \quad (1)$$

where

a_i is the measurement a taken on individual i,
a_s is the same measurement taken on the "standard" form s,
X_i is the mean of a set of measurements taken on individual i, and
X_s is the mean of the same set of measurements taken on s

(the calculation of X_i is influenced by missing data values, and in such cases the largest subset of measurements that are common to both individual and standard are used to calculate both X_i and X_s). The statistical properties of ratios are improved by converting raw measurements into logarithms (Hills, 1978). On logarithmic transformation Equation (1) becomes:

$$r_{ai} = \log(a_i) - \log(a_s) - [\log(X_i) - \log(X_s)] \quad (2)$$

Equation (2) is the transformation adopted in this study. The "standard" form is an "average primate", calculated by averaging the dimensions of a large group of specimens that included all the outgroup taxa together with the sample of modern *Homo sapiens*. We also note that Equation (2) is a generalized version of a transformation proposed by Kazmierczak (Kazmierczak, 1985; Berge & Kazmierczak, 1986). The transformed variate r_{a_i} is dimensionless and is thus independent of the scale of measurement. It expresses (in logarithmic form) the ratio between a structure measured on an individual and the same structure measured on a "standard" form, after scaling the individual and the "standard" to the same overall size. It is invariant to isometric change in overall size, and hence satisfies the criteria of Mosimann (1970) and Hills (1978) for a size-independent measure of shape. We note further advantages in this version of Kazmierczak's transformation: it is

Table 4 Character state matrix for the hominid taxa and the hypothetical outgroup (see Table 3 for identification of character symbols)

Characters	B01	B02	B03	B04	B05	B06	B07	B08	B09	B10	B11	B12	B13	B14	B15	B16	V01	V02	V03	V04	V05	V06	V07	V08	V09	V10	V11	V12	V13	F01
Outgroup	4	4	4	4	4	5	4	4	4	4	4	4	4	4	4	4	3	3	4	4	4	3	3	4	4	4	4	4	3	4
A. afarensis	4	5	7	5	9	5	3	4	4	9	9	5	9	9	9	9	6	9	4	5	9	9	9	9	4	9	4	9	9	9
A. africanus	3	4	4	4	4	4	2	4	4	4	4	4	4	3	4	3	6	3	4	4	4	4	5	4	4	4	4	4	4	3
A. robustus	9	9	9	9	9	5	4	9	4	4	9	9	5	9	9	9	6	4	9	4	9	3	9	9	9	3	9	9	9	4
A. boisei	4	5	3	5	4	5	4	4	4	4	3	4	5	5	4	3	6	3	4	4	4	3	5	2	3	3	3	3	3	4
H. erectus	4	4	4	3	4	4	4	4	4	4	4	4	3	3	3	3	7	4	5	4	5	4	5	4	5	4	6	4	5	4
H. sapiens	4	6	6	5	5	4	2	3	3	4	5	5	4	3	5	4	7	6	6	4	6	6	7	6	7	6	7	7	7	5
H. habilis (Conventional)	4	4	3	3	4	4	4	4	4	5	4	4	5	3	4	9	7	4	4	4	4	4	5	4	5	4	5	5	5	4
H. habilis (Stringer, 1986)	4	4	4	4	4	3	3	3	3	4	4	4	9	9	9	9	7	4	4	4	4	4	5	4	5	4	9	5	5	4
Homo sp. (Stringer, 1986)	4	4	3	3	4	4	4	4	4	5	9	9	4	9	4	9	7	4	4	4	5	4	5	4	5	4	5	5	5	4
H. habilis (this study)	4	4	4	4	4	3	3	3	3	4	4	4	9	9	9	9	5	4	9	4	9	9	4	9	4	9	4	9	5	9
Homo sp. (this study)	4	4	3	3	4	4	4	4	4	5	9	9	5	3	4	9	7	4	4	4	4	4	5	4	5	4	5	5	5	4

Table 4—*continued*

Characters	F 0 2	F 0 3	F 0 4	F 0 5	F 0 6	F 0 7	F 0 8	F 0 9	F 1 0	F 1 1	F 1 2	F 1 3	F 1 4	F 1 5	F 1 6	F 1 7	F 1 8	F 1 9	M 0 1	M 0 2	M 0 3	M 0 4	M 0 5	M 0 6	M 0 7	M 0 8	M 0 9	M 1 0	M 1 1	M 1 2
Outgroup	4	4	4	4	4	5	4	4	4	4	4	4	4	4	4	4	3	3	4	4	4	4	5	4	4	4	4	4	4	4
A. afarensis	9	4	9	4	9	9	9	3	9	9	9	9	9	9	9	9	9	9	3	4	4	4	4	4	6	4	4	3	3	4
A. africanus	4	5	4	4	4	4	2	5	3	4	4	4	6	4	4	4	4	1	3	4	4	5	5	4	6	3	5	3	3	4
A. robustus	9	4	4	5	4	6	9	5	9	3	4	4	6	4	4	5	5	0	4	4	4	4	6	5	6	3	9	2	3	5
A. boisei	5	5	4	4	4	5	2	4	4	4	5	5	6	4	4	5	4	0	3	4	4	6	6	5	7	3	4	2	4	5
H. erectus	4	4	4	4	4	4	3	4	3	9	4	4	5	4	3	3	3	0	3	4	4	4	5	4	6	4	5	3	3	4
H. sapiens	4	4	5	5	5	6	5	5	3	3	3	3	3	5	3	4	6	1	3	4	3	5	4	3	4	3	5	3	2	3
H. habilis (Conventional)	4	4	4	4	4	5	2	4	4	3	4	4	4	4	3	4	4	0	2	4	4	5	5	4	6	3	4	3	3	5
H. habilis (Stringer, 1986)	9	4	9	4	4	5	1	4	4	9	4	4	9	4	3	9	4	9	3	4	4	5	5	4	6	4	4	3	3	9
Homo sp. (Stringer, 1986)	9	4	5	4	4	5	2	4	3	4	4	4	4	4	3	4	4	0	2	3	4	5	5	5	6	3	5	3	3	4
H. habilis (this study)	9	9	9	4	9	9	9	9	9	9	9	9	9	9	9	9	9	9	2	3	4	5	5	4	6	4	4	3	4	4
Homo sp. (this study)	4	4	4	4	4	5	2	4	4	3	4	4	4	4	3	4	4	0	9	4	9	9	9	5	6	3	5	2	3	4

Characters	M 1 3	M 1 4	M 1 5	M 1 6	M 1 7	M 1 8	M 1 9	M 2 0	M 2 1	M 2 2	P 0 1	P 0 2	P 0 3	P 0 4	P 0 5	P 0 6	P 0 7	P 0 8	P 0 9	P 1 0	P 1 1	P 1 2	P 1 3	P 1 4	P 1 5	P 1 6	P 1 7	P 1 8	P 1 9	P 2 0
Outgroup	4	4	4	4	4	4	4	4	4	4	4	4	4	4	4	4	4	4	4	4	5	4	4	4	4	5	4	4	4	4
A. afarensis	4	4	3	3	3	5	4	5	5	5	3	4	4	3	2	3	2	2	4	5	4	3	2	3	4	4	4	4	5	5
A. africanus	3	3	3	3	3	5	5	5	4	5	4	5	4	4	5	6	[illegible]	3	4	5	9	9	9	9	4	5	4	5	6	5
A. robustus	3	2	2	2	3	5	5	5	4	5	4	5	3	4	3	4	[illegible]	3	5	5	9	9	1	3	4	5	4	5	5	5
A. boisei	2	2	1	2	3	5	4	5	5	5	3	5	3	3	5	5	[illegible]	2	5	5	3	3	1	3	5	6	4	6	5	6
H. erectus	9	3	3	3	3	5	4	5	4	4	3	4	4	4	7	5	3	4	3	4	9	9	2	3	4	4	4	4	3	4
H. sapiens	2	3	2	2	2	3	4	4	3	4	2	4	3	5	4	4	1	3	3	3	3	3	1	3	3	4	4	4	3	4
H. habilis (Conventional)	4	3	3	2	3	4	5	5	5	5	3	4	3	4	5	4	0	3	4	4	3	2	2	3	4	4	5	4	5	5
H. habilis (Stringer, 1986)	4	3	3	3	3	4	5	5	5	5	3	4	4	4	6	4	0	3	4	4	9	9	9	9	4	4	4	4	4	5
Homo sp. (Stringer, 1986)	9	9	3	3	3	4	5	5	4	4	3	4	3	4	5	5	[illegible]	3	4	4	3	2	2	3	4	4	4	4	5	4
H. habilis (this study)	4	3	3	3	3	4	5	5	5	5	3	4	4	4	9	4	9	9	4	4	9	9	9	9	4	4	4	4	4	4
Homo sp. (this study)	9	9	9	9	9	9	9	9	4	4	3	5	3	4	5	4	0	3	4	5	3	2	2	3	4	4	5	5	5	5

Note: character states are integers ranging from 0 (minimum) to 8 (maximum). A character state of 4 represents the midpoint of the range and is the value possessed by the "standard' (the "average primate"). Values of 3, 2, 1 and 0 represent successive reductions in a character relative to the standard, while values of 5 to 8 represent character enlargement relative to the standard. The value 9 indicates a missing value. The hypothetical outgroup possesses character states that are similar to, but not identical with, those of the standard.

easy to calculate and is little influenced by missing data. Provided that a reasonable number and variety of measurements can be taken on an individual then the size-correction term $\log(X_i) - \log(X_s)$ [see Equation (2)] can be estimated accurately, and the transformation can be carried out for all available measurements.

Character state determination

Currently available microcomputer phylogeny programs require character states to be defined as discrete rather than continuous values (Fink, 1986). Our transformation, however, yields continuous variables with positive and negative values [Equation (2)]. These transformed values were averaged within each taxon, and the averaged characters were then converted to discrete character states by dividing each by an empirically-determined constant (0·04), discarding fractions of a unit and then adding four units. This coding method (designated "combination coding" by Sokal & Rohlf, 1981) ensured that the character states were positive integers between the inclusive limits 0 and 8, with an average value of 4 (Table 4 gives the character states in the ingroup OTUs and an outgroup taxon). No attempt was made to standardize for the different number of character states generated for different characters (although such standardization is available as a character weighting option in the PAUP phylogenetic program). It was assumed here that characters exhibiting a narrow range of states among the ingroup taxa were evolutionarily conservative.

Cladogram construction and comparison

Cladograms for the hominidae were constructed using three different taxonomic groupings of early *Homo* (see Tables 1 and 2). Cladograms were computed for each version of the OTUs using the PAUP phylogenetic program (Swofford, 1985). In each case the branch-and-bound option was used to find the most parsimonious unrooted tree, and the trees were rooted to form cladograms by including a hypothetical outgroup in the study. This outgroup was taken as being identical with the shared morphotype (i.e., shared derived characters) or *Pan, Gorilla* and *Pongo*. The hypothetical outgroup's character states were identified by imposing a "molecular" phylogeny on the six outgroup taxa (extant primates) and then finding the most parsimonious interpretation of the great ape morphotype. This procedure is referred to as "optimization" (Swofford, 1985), and we used the PAUP program's "MINF" algorithm to assign optimal ancestral character states to the great ape morphotype. The molecular cladogram for the outgroup taxa considered here is given in Figure 1.

The use of a single morphotypic outgroup is advantageous for two reasons. Firstly, the homoplasy that is present among the outgroup taxa is not allowed to make a direct contribution to the maximum parsimony solution for the ingroup taxa. This is important because the characters used in this study were selected for their potential phyletic valence

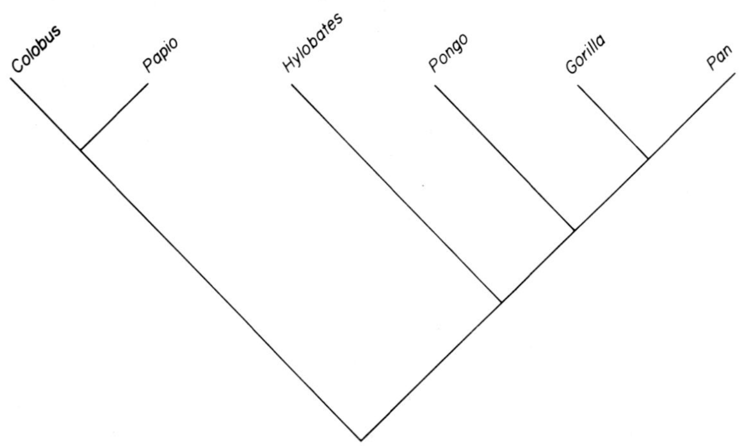

Figure 1. Cladogram of outgroup taxa based on molecular evidence.

among hominid taxa, and they do not necessarily evolve parsimoniously among the taxa of a wider group. Secondly, the ability of programs such as PAUP to find the most parsimonious tree for a set of taxa is greatly facilitated by keeping the number of taxa to a minimum.

We selected the great ape morphotype, rather than the African apes, as a hypothetical outgroup because the branching sequence among *Pan, Gorilla* and *Homo* is still unresolved (Groves, 1986; Savatier *et al.*, 1987). In view of the possibility that *Pan* and *Gorilla* share a common ancestor that post-dates the origin of the hominid radiation, it was in the authors' view more prudent to select a morphotype which virtually all workers agree as being ancestral to the hominid lineage.

Results

Cladograms: conventional OTUs

The most parsimonious cladogram for conventional OTUs is shown in Figure 2. This cladogram is consistent with the phylogeny advanced by Kimbel *et al.* (1984). Our data also supports alternative cladograms which are only slightly less parsimonious than the one shown in Figure 2. In these less parsimonious arrangements of the conventional OTUs, *A. africanus* appears as the sister group of *Homo*, or as the sister group of a *Homo*/"robust" australopithecine clade, while the positions of the remaining taxa are unchanged.

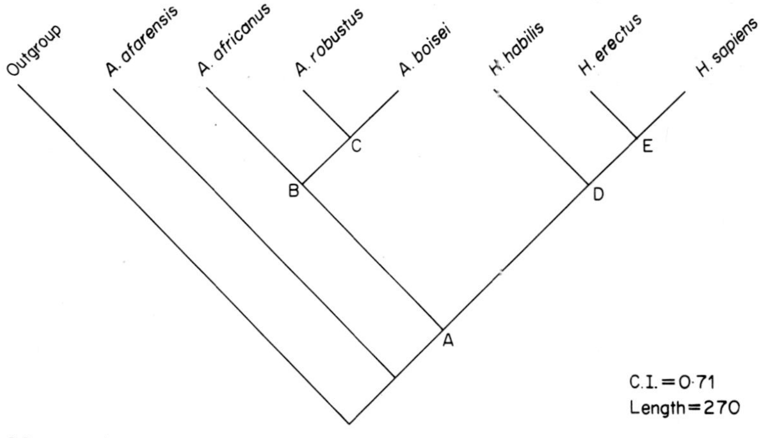

Figure 2. Most parsimonious cladogram for conventional hominid taxa. C.I. = Consistency Index, Length = No. of Evolutionary Steps (character state changes). Shared derived character states are not listed for the ingroup node (branching point for *A. afarensis*) since these character states are determined by the choice of outgroup.

Shared derived character states for labelled nodes
(see Table 3 for identification of character symbols)

A. M4, M5, *M8, M13, M14, M16, P3, P5, P7, P11*
B. $F14^2$, P2, *P7,* P10, P16, P18
C. B6, B13, *V6, V10,* F17, M5, M6, *M10,* M12, *M14, M15,* P9, *P13*
D. V1, V2, V9, V11, V13, *F11, F16*
E. V3, V5, V11, F8, *F10,* M9, *M22, P9,* $P19^2$, *P20*

Key
F3 .... Increase in character F3 by one character state
$F3^2$ .... Increase in character F3 by two character states
F3 .... Decrease in character F3 by one character state
$F3^2$ .... Decrease in character F3 by two character states

Subdivisions of early Homo

The most parsimonious cladogram using Stringer's subdivision of early *Homo* is given in Figure 3. The two species of early *Homo* (*Homo habilis* and *Homo* sp.) appear on this cladogram as successive sister taxa to a clade comprising *H. erectus* and *H. sapiens*. In other respects the cladogram is identical with the cladogram containing *H. habilis sensu lato* (Figure 2). Slightly less parsimonious alternatives to the cladogram shown in Figure 3 exist: in essence, they allow an interchanging of the relationships among the four species in the "*Homo*" clade, and they also allow *A. africanus* to be placed as the sister taxon to *Homo* rather than to the "robust" australopithecines.

The most parsimonious cladogram using Chamberlain and Wood's subdivision of early *Homo* (Figure 4) differs radically from the other cladograms. Chamberlain and Wood's *Homo* sp. is most parsimoniously interpreted as the sister taxon of the "robust" australopithecines, with *A. africanus* as the sister taxon of this clade. *H. habilis*, on the other hand, appears to be a relatively primitive hominid that nonetheless shares some derived characters with other post-*A. afarensis* hominid taxa. Slightly less parsimonious reconstructions allow different sister-group relationships between (but not within) four basic groups: *A. afarensis*, *H. habilis*, and *H. erectus*/*H. sapiens* clade and the *A. africanus*/*Homo* sp./"robust" australopithecine clade.

Discussion

Our intention in this study was to explore the possibility of extracting phylogenetic information from the pattern of shape variation among the various species of hominid. This analysis is an example of "low-level taxonomy" (Arnold, 1981) and we refer the reader to

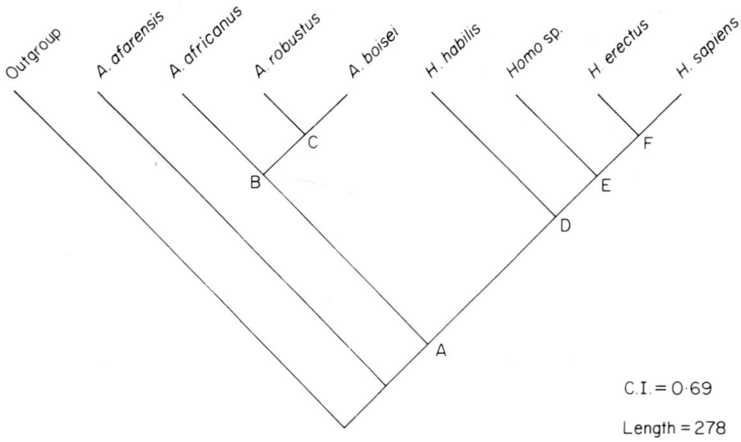

Figure 3. Most parsimonious cladogram for Stringer's subdivision of early *Homo*.

Shared derived character states for labelled nodes
(see Table 3 for identification of character symbols)

A. M4, M5, *M8, M13, M14,* M19, P5, P6, *P7², P11*
B. F3, F9, F14², P1, P2, *P13*, P16, P18
C. B2, *B3*, B4, B6, B7, *B11*, B13, B14², *V6, V8², V9, V10, V11, V12, V13,* F2, F17, M5, M6, *M10*, M12, *M14, M15, M16, P3*, P9
D. V1, V2, V9, V11, V12, V13, F4, *F16, M18, P10, P19*
E. *B4*, B7, V5, *F10*, M9, *M13, M22, P3,* P7, *P20*
F. V3, V11, F8, *F11, M19, P9, P19*
(Key as for Figure 2).

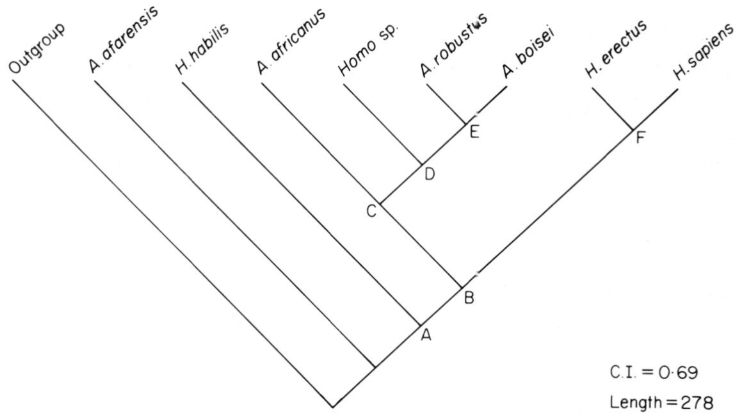

Figure 4. Most parsimonious cladogram for Chamberlain and Wood's subdivision of early *Homo*.
Shared derived character states for labelled nodes
(see Table 3 for identification of character symbols)

A. M4, M5, *M14*, M19, P5, *P7, P11*
B. V7, *M8*, M9, *M13, M21*
C. *V13, F8*, F14, P2, *P7*, P10, P16, P18, P19, P20
D. *B3*, B7, *B11*, B13, F7, F10, *F19*, M5, M6, *M10, M14, M15, M16, P3*
E. B2, B4, B6, $B14^2$, *V6*, *$V8^2$*, *V9, V10, V11, V12, V13*, F2, F17, *M9*, M12, P9, *P13*
F. V1, V3, V5, V9, $V11^2$, *F16, M13, M19, M22, P9, P19*
(Key as for Figure 2).

the latter author's guidelines for work of this nature. We have followed Arnold's recommendations in using a large character base sampled across different anatomical regions, in including all of the main hominid taxa and in sampling the sexual and geographic variation in those taxa. Arnold (1981) also noted that in attempting phylogenetic reconstruction at the subfamily or genus level the problem of the composition of the OTUs is magnified. The results of this study illustrate the potential influence on hominid phylogeny of changing the composition of one of the OTUs. However, sister group relationships of some of the taxa remained stable despite making such changes. The "robust" australopithecines are consistently grouped together, and the later species of *Homo* (*H. erectus* and *H. sapiens*) also retain a close cladistic relationship.

As in our previous study (Wood & Chamberlain, 1985) we have used outgroup analysis as our sole means of determining the polarity of character state changes. However, rather than assuming that "common-is-primitive" among the outgroup we have used our knowledge of outgroup structure to refine our estimate of ancestral character states (see Maddison *et al.*, 1984, for a justification of this strategy). The hypothetical outgroup is assumed to be primitive in all characters relative to the hominids, and it thus allows the unrooted maximum parsimony trees to be rooted by placing the outgroup as the sister group of all other taxa. Ontogenetic information, perhaps the most reliable indicator of character polarity (Nelson, 1978; Bonde, 1984), was not employed here because it is much less readily available for the characters used in this study. Juvenile specimens tend to be poorly preserved in the hominid fossil record, and are less reliably assigned to species. Furthermore, interpretations of character evolution based on ontogeny are confounded by paedomorphosis, a phenomenon that is often cited as having influenced the later stages of human ancestry. Functional interpretations of character evolution are favoured by some authors, but we share Bonde's (1977) unease at such an approach.

Comparisons with other cladistic studies

In an earlier study, which was based on a different set of data (Wood and Chamberlain, 1986), the characters of the vault and base supported a close phylogenetic relationship between *A. afarensis* and the "robust" australopithecines (see Figures 2 and 5 in Wood & Chamberlain, 1986), while the characters of the face, palate and mandible supported an *A. africanus*/"robust" clade (see their Figures 3, 4 and 6). In the earlier study, the overall cladograms favoured the vault/base pattern of relationships. In the present study, with its greater emphasis on dental and gnathic traits, the maximum parsimony cladograms achieve consensus in indicating that *A. afarensis* is quite primitive while *A. africanus* has affinities with the "robust" australopithecines. As in the earlier study, the cladistic relationships of *A. africanus* are by no means certain since this species can be placed as the sister group of *Homo* with only a small increase in homoplasy.

The present analysis does not show a particularly close, or exclusive, relationship between the "robust" australopithecines and early *Homo* as conventionally defined. However, *Homo* sp. as defined by our phenetic analysis does show some cladistic relationship with the "robust" taxa (Figure 4). Other workers have proposed that the "robust" australopithecines are closely related to early *Homo*. Dean (1986) lists derived characters (but see Grine, 1987) of the basicranium and of the developing dentition that are common to both groups, and Skelton *et al.* (1986), in a compatibility analysis using a wider range of characters, also found synapomorphies linking the two groups. In the present study we suspect that certain characters of the base, face and mandible in particular indicate a close phylogenetic relationship between *Homo* sp. (*sensu* Chamberlain and Wood) and the "robust" australopithecine clade. However, this relationship does not appear to extend to other species of *Homo*.

Another body of evidence suggests that the "robust" australopithecines share a most recent common ancestor with *A. afarensis*. Olson (1981, 1985) has cited derived characters of the basicranium and nasal region in support of this pattern of relationships, and Wood & Chamberlain (1986) also offered tentative evidence for this cladistic hypothesis. In the present study, in which proportionately more of the characters were recorded on the face, palate and mandible, this particular cladistic hypothesis does not receive support.

The effect of subdividing early Homo

Stringer's subdivision of early *Homo* was based primarily on size and only secondarily on shape considerations (Stringer, 1986, pp. 287–288). Since size was removed *ab initio* in the present study, it is not surprising that Stringer's taxa are not widely separated on the cladograms depicted here. Stringer's subdivision of early *Homo* cuts across the subdivision advocated in the present study (Table 2), thus providing some indication of the orthogonality of the "shape" criteria on which our phenetic subdivision is founded. The two early *Homo* OTUs that we have proposed behave very differently under maximum parsimony analysis. Our *Homo* sp. has affinities (or "resistant" homoplasies) which align it with the "robust" taxa: such affinities appear to involve characters of the base, face and mandible. Our version of *H. habilis*, on the other hand, shares mandibular and palatal characters with other post-*A. afarensis* taxa, but is in other respects a "primitive" hominid.

Conclusions

In this analysis we have aimed at achieving a more reliable reconstruction of cladistic relationships among hominids through improvements to methods used in an earlier study (Wood & Chamberlain, 1986). We follow Arnold (1981) in holding that the reliability of any cladistic analysis is likely to be influenced by a number of factors including character selection and weighting, OTU composition, choice of computational criteria (e.g., compatibility methods *versus* maximum parsimony), polarity determination, etc., and we view these potential confounding factors as a challenge rather than as an indictment of the cladistic approach. We see the need to improve our analysis by including more characters from a wider range of anatomical regions, by including ontogenetic information and perhaps by weighting characters according to their within-group variation. Ultimately, we aim to construct cladograms which can be considered "reliable" according to the following criteria. Firstly, the cladogram should be internally consistent, as judged by measures of homoplasy such as the Consistency Index, or by the stability of its topology on selection of subsamples of the character set. Secondly, the cladogram should be consistent with other cladistic analyses, particularly where the latter have been based on independent sets of characters or have analysed in greater detail the affinities of smaller groups of OTUs. Thirdly, the most powerful test of any cladogram subsists in the predictions that it supports. These are several, and concern the nature of ancestral forms, the stratigraphic order of appearance of different lineages, palaeobiogeography, ontogeny and so forth. Future fossil discoveries and more detailed study of the existing evidence provide the means for testing such predictions.

This cladistic study only satisfies in part the first two criteria. Our cladograms exhibit sufficient homoplasy to allow alternative cladistic interpretations that are only slightly less parsimonious than the trees given here. They do, however, agree with a number of other cladistic studies in identifying both the "robust" australopithecines and the later species of *Homo* as pairs of sister taxa. The most intriguing aspect of this analysis is our proposal that a taxonomic subdivision of early *Homo* can be made that identifies morphs that are geographically, as well as cladistically, distinct. We await detailed analysis of new fossil hominid material from Olduvai Gorge, Tanzania, with renewed interest.

Acknowledgements

We are grateful to the Director and Trustees of the National Museums of Kenya and the Government of Tanzania for allowing us to study fossils in their care. We also thank the Director and Trustees of the Transvaal Museum and the Head of the Department of Anatomy, University of the Witwatersrand, for giving us permission to examine fossils for which they are responsible. The staff of the British Museum (Natural History) and of the Powell Cotton Museum kindly allowed access to their collections of primate skeletal material. This research was facilitated by funds from the Natural Environment Research Council, U.K. and The Leverhulme Trust, and was made possible by an invitation from Richard Leakey to undertake a long-term study of part of the hominid collection from Koobi Fora, Kenya.

References

Arnold, E. N. (1981). Estimating phylogenies at low taxonomic levels. *Z. zool. Syst. Evolut.-forsch.* **19,** 1–35.

Atchley, W. R., Gaskins, C. T. & Anderson, D. (1976). Statistical properties of ratios. 1. Empirical results. *Syst. Zool.* **25,** 137–148.

Berge, C. & Kazmierczak, J. B. (1986). Effects of size and locomotor adaptations on the hominid pelvis: evaluation of australopithecine bipedality with a new multivariate method. *Folia primat.* **46,** 185–204.

Bonde, N. (1977). Cladistic classification as applied to vertebrates. In (M. K. Hecht *et al.*, Eds) *Major Patterns in Vertebrate Evolution*, pp. 741–804. New York: Plenum.

Bonde N. (1984). Primitive features and ontogeny in phylogenetic reconstructions. *Vidensk. Meddr dansk naturh. Foren.* **145,** 219–236.

Chamberlain, A. T. (in preparation). A taxonomic review and phylogenetic analysis of *Homo habilis*. Ph.D. Thesis, University of Liverpool.

Charig, A. J. (1982). Systematics in biology: a fundamental comparison of some major schools of thought. In (K. A. Joysey & A. E. Friday, Eds) *Problems of Phylogenetic Reconstruction*, pp. 363–440. London: Academic Press.

Corruccini, R. S. & McHenry, H. M. (1980). Cladometric analysis of Pliocene hominids. *J. hum. Evol.* **9,** 209–221.

Day, M. H. (1986). *Guide to Fossil Man.* 4th edition. London: Cassel.

Dean, M. C. (1986). *Homo* and *Paranthropus*. In (B. A. Wood *et al.*, Eds) *Major Topics in Primate and Human Evolution*, pp. 250–266. Cambridge: Cambridge University Press.

Delson, E. (1986). Human phylogeny revised again. *Nature, Lond.* **322,** 496–497.

Delson, E., Eldredge, N. & Tattersall, I. (1977). Reconstruction of hominid phylogeny: a testable framework based on cladistic analysis. *J. hum. Evol.* **6,** 263–278.

Eldredge, N. & Tattersall, I. (1975). Evolutionary models, phylogenetic reconstruction, and another look at hominid phylogeny. In (F. S. Szalay, Ed.) *Approaches to Primate Paleobiology*, pp. 218–242. Basel: Karger.

Fink, W. L. (1986). Microcomputers and phylogenetic analysis. *Science, N.Y.* **234,** 1135–1139.

Grine, F. E. (1987). On the eruption pattern of the permanent incisors and first permanent molars in *Paranthropus*. *Am J. phys. Anthrop.* **72,** 353–359.

Groves, C. P. (1986). Systematics of the great apes. In (D. R. Swindler & J. Erwin, Eds) *Comparative Primate Biology*. Vol. 1, *Systematics, Evolution, and Anatomy*, pp. 187–217. New York: Liss.

Hills, M. (1978). On ratios—a response to Atchley, Gaskins, and Anderson. *Syst. Zool.* **27,** 61–62.

Howell, F. C. (1978). Hominidae. In (V. J. Maglio & H. B. S. Cooke, Eds) *Evolution of African Mammals*, pp. 154–248. Cambridge: Harvard University Press.

Hull, D. L. (1984). Cladistic theory: hypotheses that blur and grow. In (T. Duncan & T. F. Stuessy, Eds) *Cladistics: Perspectives on the Reconstruction of Evolutionary History*, pp. 5–23. New York: Columbia University Press.

Johanson, D. C. & White, T. D. (1979). A systematic assessment of early African hominids. *Science, N.Y.* **202,** 321–330.

Johanson, D. C., Masao, F. T., Eck, G. G. *et al.* (1987). New partial skeleton of *Homo habilis* from Olduvai Gorge, Tanzania. *Nature, Lond.* **327,** 205–209.

Kazmierczak, J. B. (1985). Analyse logarithmique. Deux exemples d'application. *Revue Statist. appl.* **33,** 13–24.

Kennedy, G. E. (1983). Femoral morphology of *Homo erectus*. *J. hum. Evol.* **12,** 587–616.

Kimbel, W. H., White, T. D. & Johanson, D. C. (1984). Cranial morphology of *Australopithecus afarensis*: a comparative study based on a composite reconstruction of the adult skull. *Am. J. phys. Anthrop.* **64,** 337–388.

Maddison, W. P., Donoghue, M. J. & Maddison, D. R. (1984). Outgroup analysis and polarity. *Syst. Zool.* **33,** 83–103.

Mosimann, J. E. (1970). Size allometry: size and shape variables with characterisations of the lognormal and generalised gamma distributions. *J. Am. statist. Ass.* **65,** 930–945.

Nelson, G. (1978). Ontogeny, phylogeny, paleontology, and the biogenetic law. *Syst. Zool.* **27,** 324–345.

Olson, T. R. (1981). Basicranial morphology of the extant hominoids and Pliocene hominids: the new material from the Hadar Formation, Ethiopia, and its significance in early human evolution and taxonomy. In (C. B. Stringer, Ed.) *Aspects of Human Evolution*, pp. 99–128. London: Taylor and Francis.

Olson, T. R. (1985). Cranial morphology and systematics of the Hadar Formation hominids and "*Australopithecus*" *africanus*. In (E. Delson, Ed.) *Ancestors: the Hard Evidence*, pp. 102–119. New York: Liss.

Savatier, P., Trabuchet, G., Chebloune, Y. *et al.* (1987). Nucleotide sequence of the beta globin genes in Gorilla and Macaque: the origin of nucleotide polymorphisms in human. *J. molec. Evol.* **24,** 309–318.

Skelton, R. R., McHenry, H. M. & Drawhorn, G. R. (1986). Phylogenetic analysis of early hominids. *Curr. Anthrop.* **27,** 21–43.

Sokal, R. R. & Rohlf, F. J. (1981). *Biometry: the Principles and Practice of Statistics in Biological Research,* 2nd edition. New York: Freeman.

Stringer, C. B. (1986). The credibility of *Homo habilis*. In (B. A. Wood *et al.*, Eds) *Major Topics in Primate and Human Evolution*, pp. 266–294. Cambridge: Cambridge University Press.

Swofford, D. R. (1985). *PAUP. Phylogenetic Analysis Using Parsimony. Version 2.4.* Champaign: Illinois Natural History Survey.
Tattersall, I. & Eldredge, N. (1977). Fact, theory, and fantasy in human paleontology. *Am. Scient.* **65,** 204–211.
Tobias, P. V. (1980). "*Australopithecus afarensis*" and *A. africanus*: critique and an alternative hypothesis. *Palaeont. afr.* **23,** 1–17.
Trinkaus, E. (1984). Does KNM-ER 1481A establish *Homo erectus* at 2·0 myr BP? *Am. J. phys. Anthrop.* **64,** 137–139.
Walker, A., Leakey, R. E., Harris, J. M. & Brown, F. H. (1986). 2·5-Myr *Australopithecus boisei* from west of Lake Turkana, Kenya. *Nature, Lond.* **322,** 517–522.
White, T. D., Johanson, D. C. & Kimbel, W. H. (1981). *Australopithecus africanus*: its phyletic position reconsidered. *S. Afr. J. Sci.* **77,** 445–470.
Wood, B. A. & Chamberlain, A. T. (1986). *Australopithecus*: grade or clade? In (B. A. Wood *et al.*, Eds) *Major Topics in Primate and Human Evolution*, pp. 220–248. Cambridge: Cambridge University Press.

C. B. Stringer

Department of Palaeontology, British Museum (Natural History), Cromwell Road, London SW7 5BD, England

Received 2 March 1987
Revision received 12 June 1987
and accepted 17 June 1987
Publication date November 1987

Keywords: Cladistic analysis, PAUP, *Homo habilis, Homo erectus,* Neanderthals, *Homo sapiens.*

A numerical cladistic analysis for the genus *Homo*

Using the PAUP (phylogenetic analysis using parsimony) program, nine different hominid groups allocated to the genus *Homo* have been included in a numerical cladistic analysis. Eleven (essentially continuous) characters, mainly cranial, have been used for the analysis, and the most parsimonious trees generated by PAUP have been compared with a "conventional" tree provided by the author. PAUP's three best trees all assign the early archaic *H. sapiens* group to a clade containing early African *H. erectus* and Asian *H. erectus*, rather than to the clade containing Neanderthals, African late archaic *H. sapiens*, Skhūl-Qafzeh and modern *H. sapiens*. Relationships within the latter clade could not be resolved clearly by PAUP, although the Skhūl-Qafzeh and modern *H. sapiens* groups were consistently linked through synapomorphies. As long as ordered characters were used, the two subsets of *H. habilis* were treated as plesiomorphous, while modern *H. sapiens* was highly derived, suggesting that characters in common were homoplasies.

Journal of Human Evolution (1987) **16,** 135–146

Introduction

It is only in the last decade that cladistic methods of analysis have made an impact on reconstructions of hominid phylogeny, but a survey of recent publications demonstrates that the advantages of such methods are now becoming widely appreciated (see, for example, Andrews & Franzen, 1984; Delson, 1985; Wood *et al.*, 1986).

Early attempts at phylogenetic analyses of hominids were wide-ranging and very simple, but nevertheless they challenged some of the conventional wisdom about human evolution and in some cases raised issues which have not yet been resolved. Thus Eldredge & Tattersall (1975), while accepting that "*Ramapithecus*" was an early hominid on cladistic grounds and that *Australopithecus africanus* was essentially primitive, raised the problem that *Homo habilis* and (modern) *H. sapiens* both lacked derived characters found in the supposedly "intermediate" species *H. erectus*. Additionally, they noted that the Neanderthals appeared to share many of the derived characters of *H. erectus*. These views were reiterated in Tattersall & Eldredge (1977), but a more conventional view was taken by Delson *et al.* (1977), where *H. habilis* was viewed as the sister group of *H. erectus* and *H. sapiens* (which included the Neanderthals). At about the same time, Bonde (1977) attempted a classification of the hominids using cladistic methods, and utilized some conventions for dealing with extinct taxa and possible ancestors, which were further employed in Bonde (1981). Olson (1978) produced a cladistic definition of the genus *Homo*, including within it *Homo (Australopithecus) africanus* (uniting specimens more usually attributed to *A. africanus* and *H. habilis*), and the additional species *erectus, neanderthalensis* and *sapiens*. Olson also expressed doubt about the phylogenetic position of the Neanderthals, in that their sister group might be either *H. erectus* or *H. sapiens*.

Doubts about the phylogenetic position of *H. erectus* have been raised again recently after a cladistic analysis by Andrews (1984), using data provided by several other workers. Andrews argued that rather than postulate a series of reversals by which the apparent autapomorphies of *H. erectus* were lost with the appearance of *H. sapiens*, it was more

0047–2484/87/010135 + 12 $03.00/0

parsimonious to argue that any evolutionary transition from an *H. habilis*-like ancestor to *H. sapiens* did not pass through a stage represented by Asian *H. erectus*. Furthermore, he provided a list of possible synapomorphies for the *H. habilis*/modern *H. sapiens* clade. In the same volume, Stringer (1984*a*) and Wood (1984), while not taking exactly the position of Andrews, noted the apparent autapomorphies of Asian *H. erectus* when compared with the more plesiomorphous condition found in early African specimens usually referred to the same species.

Several workers (e.g., Bräuer, 1984; Howells, 1980; Rightmire, 1984; Stringer, 1984*a*, 1985) have argued for the existence of a Middle Pleistocene hominid group separate from Asian *H. erectus*, and sharing derived characteristics with Upper Pleistocene hominids conventionally classified together as *H. sapiens*. This group is generally called "archaic" *H. sapiens*, although the unsatisfactory nature of this blanket term is recognized (Stringer, 1984*b*, 1985; Tattersall, 1986). Among the range of hominids usually assigned to "archaic" *H. sapiens*, there are a number of specimens sharing apparent synapomorphies with *H. erectus*, and the classification of these specimens (e.g., Bodo, Arago and Petralona) continues to be a vexing question (Stringer, 1983, 1984*a*, 1985; Hublin, 1985, 1986). Equally, the phylogenetic position of the Neanderthals still remains a problem (with workers such as Bräuer, 1984, and Stringer, 1985, arguing that modern *H. sapiens* derived from African late archaic *H. sapiens* rather than from late Neanderthals). However, Upper Pleistocene human evolution has so far received less attention from workers employing cladistic techniques, perhaps partly because of the difficulty of treating possible subspecific variation cladistically.

A new cladistic analysis for the genus *Homo*

Having gained access to the PAUP (Phylogenetic Analysis Using Parsimony) package, it was decided to attempt a relatively simple cladistic analysis using this program. PAUP is considered to be "the most impressive package currently available" for maximum parsimony analysis (Fink, 1986, p. 1136). It relies on the Wagner criterion for tree construction (Wiley, 1981; Swofford, 1985) and provides a number of options for character assignments, character ordering and tree production. Tree length and consistency index (a relative measure of the economy of the tree produced, as defined by Kluge & Farris, 1969) can be compared between trees, and it is possible to supply a chosen topology for testing. Information is also available on character changes and synapomorphies through the trees generated.

A fairly standard character set of 11 skeletal characters was produced for nine human groups (Table 1) and analysed by PAUP to investigate the following problems:

(i) What would be the form of tree(s) deemed most parsimonious by PAUP?
(ii) If the most parsimonious tree(s) produced by PAUP did not match with a conventional time-based phyletic sequence for the genus *Homo*, what was the nature of the differences, given that a conventional sequence could be provided for comparison?
(iii) Using parsimony analysis, what was the phylogenetic relationship between *H. habilis, H. erectus* and modern *H. sapiens*?
(iv) Would the early African *H. erectus* group form the sister group to the Asian *H. erectus* group?

(v) What was the phylogenetic position of the early archaic *H. sapiens* group?
(vi) What was the phylogenetic relationship of Neanderthals, African late archaic *H. sapiens* and modern *H. sapiens*?

Groups analysed (see Table 1)
Specimens included in the hominid groups for character state assignments were as follows:

(1) *H. habilis* A (HABA): O.H. 7 and 24; KNM-ER 1470, 1472, 1481, 1590, 1802, 3228 and 3732 (Stringer, 1986).
(2) *H. habilis* B (HABB): O.H. 13 and 16; KNM-ER 1805 and 1813 (excluding the problematic KNM-ER 992 mandible, but otherwise following Stringer, 1986, in distinguishing a subset of *H. habilis* on the basis of endocranial volume, face shape, cranial morphology and overall size).
(3) Early African *H. erectus* (EAFE): O.H. 9; KNM-ER 3733, 3883; WT-15000 (Rightmire, 1984; Stringer, 1984*a*; Wood 1984; Brown *et al.*, 1985).
(4) Asian *H. erectus* (ASE): Indonesian *H. erectus* specimens (excluding the Ngandong sample); Zhoukoudian Lower Cave (Rightmire, 1984; Stringer, 1984*a*).
(5) Early archaic *H. sapiens* (EAS): Arago sample; Petralona; Bodo 1 and 2; Broken Hill sample (Stringer, 1984*a*).
(6) Neanderthals (NEA): European and Asian Neanderthals (Stringer *et al.*, 1984; Trinkaus, 1983, 1984*b*).
(7) African late archaic *H. sapiens* (AAS): Irhoud sample; Laetoli hominid 18; Florisbad; and Eliye Springs KNM-ES 11693, included on the grounds of morphology, although undated (Bräuer, 1984; Bräuer & Leakey, 1986).
(8) Skhūl-Qafzeh (SQ): Middle Palaeolithic-associated samples (Vandermeersch, 1981; Trinkaus, 1983, 1984*b*).
(9) Modern humans (MOD): an "average" morphology (Howells, 1973; Stringer 1984).

Characters used (see Table 1)
PAUP requires coding for characters in the form of values such as 0/1 for absence/presence data or from 0 up to 9 for multiple state or continuous data (except where one of the codes is used for missing data). The characters chosen here are mainly traditional ones which can be represented by metrical data, and all are essentially continuous by nature. Presence/absence characters do exist for hominids, but most appear to be autapomorphies present in only one of the groups (e.g., the hyperrobust tympanic of Asian *H. erectus* or the suprainiac fossa of Neanderthals), so it was not considered useful to include them in these preliminary analyses. Instead, the range of particular hominid morphologies has been assessed, extreme values scored from 0–2, 0–3 or 0–4, and intermediate morphologies have been allocated intermediate scores as appropriate.

For six characters (occipital angulation, midface projection, NLH/EKB ratio, facial prognathism, occipital proportions and brain size) a straightforward use of one or two measurements, ratios or angles allowed the groups to be ordered in a series by mean values and allocated character scores. For the other five characters a more intuitive, and therefore more subjective, ordering of the groups was employed based on the author's assessment of the overall morphology displayed. This was considered preferable at this stage to scoring a large number of related features separately (e.g., those reflecting cranial robusticity), since

Table 1 **Character assignments for hominid groups (U = unknown)**

Groups	Cranial robust. CROB	Postcr. robust. PROB	Supr. torus BROW	Cranial flatt. CRFL	Occip. ang. OCAN	Midface proj. MFPJ	High NLH/EKB N/E	Reduced progn. REPN	Low LI/IO REOH	Large brain BRN	Dental red. REDN
Inferred ancestor	0	1	1	1	0	1	1	0	1	0	0
H. habilis A HABA	0	1	1	1	0	0	1	1	1	1	1
H. habilis B HABB	0	U	1	1	1	1	1	2	1	1	2
Early African *erectus* EAFE	1	1	2	2	2	1	1	2	1	2	3
Asian *erectus* ASE	3	2	3	2	3	1	0	2	2	2	3
Early archaic *sapiens* EAS	2	2	3	1	3	2	1	3	1	3	3
Neanderthals NEA	1	1	2	1	1	3	2	3	0	4	3
African archaic *sapiens* AAS	1	U	2	1	1	1	0	3	0	3	3
Skhūl-Qafzeh SQ	0	0	1	0	0	1	0	3	0	4	3
Modern *sapiens* MOD	0	0	0	0	0	2	1	4	0	3	4

it can be argued that the general level of skeletal hypertrophy differences may be related to simple developmental factors (Hublin, 1986).

Selection of an outgroup for the analysis is complicated by the fact that there is no agreement as to which hominid represents the sister to the genus *Homo*. Employing the most widely used taxon names, on the assumption that *Australopithecus afarensis* was the last common ancestor for *A. africanus, A. robustus, A. boisei* and *Homo* (Kimbel *et al.*, 1984), the characters of *A. afarensis* could be used for the outgroup. However, other workers have proposed recently that *A. africanus* (*H. africanus* for Olson, 1985) or *A. robustus/boisei* represent the sister group for *Homo* (Olson, 1985; Dean, 1986; Skelton *et al.*, 1986). Because of these disagreements, I have constructed a hypothetical outgroup character set from the morphologies present in the whole "australopithecine" group, and refer to other hominoids (living "great" and "small" apes) where there are problems in resolving polarities (see "Inferred ancestor" of Table 1).

(1) *Cranial robusticity (CROB)*. Ranging from the thin unbuttressed crania of *H. habilis* A and B, and the Skhūl-Qafzeh and modern crania (scored with the outgroup as 0) to the thick, heavily buttressed crania typical of Asian *H. erectus* (scored as 3) (for discussion of this character see Andrews, 1984; Rightmire, 1984; Stringer, 1984*a*; Wood, 1984). The assumption is made here that the gracile crania of *H. habilis* and modern *H. sapiens* can be considered homologous, as Andrews, 1984, but the analysis is intended to test the validity of this assumption).

(2) *Postcranial robusticity (PROB)*. Ranging from the gracile postcranial skeleton of the Skhūl-Qafzeh and modern groups (scored 0) to the heavily built postcrania found in Asian *H. erectus* and archaic *H. sapiens* (scored 2). Comparatively assessing the outgroup morphology is difficult because of differences in overall anatomy and body size between "australopithecines" and members of the genus *Homo*. Allowing for smaller body size in "australopithecines", an outgroup score of 1 has been selected for the PAUP analyses. The *H. habilis* B and African late archaic *H. sapiens* groups have been assigned no scores because of the lack of well associated postcranial material (for data used to score groups see Trinkaus 1983, 1984*a,b*).

(3) *Supraorbital torus morphology (BROW)*. Ranging from the minimal torus development of living humans (scored 0), to the evenly thick supraorbital torus development found in Asian *erectus* and early archaic *H. sapiens* crania (scored 3). Outgroup morphology is unlikely to match either of the extreme conditions, and is scored as 1 (for discussion see Andrews, 1984; Rightmire, 1984; Stringer, 1984*a*; Wood, 1984).

(4) *Cranial flattening and elongation (CRFL)*. Ranging from the short, high and rounded cranial shape found in the Skhūl-Qafzeh and modern human groups (scored 0) to the low and platycephalic cranial form of Asian *H. erectus* (scored 2). The cranial shape indices of the early African *H. erectus* sample are insufficiently distinct from those of the Asian form to warrant a different score, but the enlarged cranial height and expanded parietal form of early archaic *H. sapiens* justify a score of 1 for that group The outgroup condition can also be scored as 1 (for discussion see Andrews, 1984; Rightmire, 1984; Stringer, 1984*a*; Wood, 1984).

(5) *Occipital angulation (OCAN)*. Arguably this feature is closely linked with cranial robusticity and occipital proportions, but given the different combinations shown in the

character sets, it has been retained as a separate character. Ranging from the rounded occipital profile of *H. habilis* A and the Skhūl-Qafzeh and modern groups (scored 0) to the extreme angulation found in the Asian *H. erectus* and early archaic *H. sapiens* specimens (scored 3). Where strong nuchal cresting does not modify the occipital profile in early hominids and other hominoids, it is relatively rounded, suggesting an outgroup score of 0 or 1. For PAUP, a score of 0 has been used (for discussion see Andrews, 1984; Hublin, 1986; Rightmire, 1984; Stringer, 1984*a*; Wood, 1984).

(6) *Midfacial projection (MFPJ)*. Ranging from the transversely flattened midface of *H. habilis* A (scored 0) to the highly projecting midface of Neanderthals (scored 3). Scoring the outgroup condition is difficult since apes (and *A. afarensis*?) display flattened upper faces (high nasiofrontal angle values) but projecting lower faces (low subspinale angle values). Other "australopithecines" have similar or even more extreme values for facial flattening than *H. habilis* A (i.e. both angle values very high), suggesting an outgroup score of 0. On the uncertain assumption that this was *not* the ancestral condition for *Homo*, I have scored the outgroup as 1, while recognizing that the condition was probably different in detail from that of *H. habilis* B (also scored as 1). For discussion of hominoid variation see Stringer (1986).

(7) *NLH/EKB index (N/E)*. A number of differences in facial proportions occur in the genus *Homo*, and data for the ratio of nasal height (NLH) to biorbital breadth (EKB) are readily available as an example. Low values of this index are found in Asian *H. erectus* (or at least in Sangiran 17!), African late archaic *H. sapiens* and the Skhūl-Qafzeh group (all scored 0), compared to the very high values found in Neanderthals (scored 2). Assessing the outgroup condition is very difficult since many "australopithecines" have intermediate values of the index, but those with very long faces (such as O.H.5) have even higher values than Neanderthals. Looking further afield to other hominoids shows a similar pattern, with long faced apes having very high values for the index, and short faced forms (gibbons) having low values. For PAUP the outgroup condition has been scored at 1 (the use of this index for later hominids is discussed in Stringer & Trinkaus, 1981, and Stringer, 1984*a*).

(8) *Reduced facial prognathism (REPN)*. The extent to which the face is retracted under the cranial vault can be measured in various ways, but here the nasion and prosthion angle values have been used to assign character states. These range from the highly prognathic condition found in *H. habilis* A (scored 1) to the orthognathic condition found in modern humans (scored 4). "Australopithecine" data all suggest an outgroup condition more extreme than that of *H. habilis* A, so this has been scored as 0 (for discussion see Stringer, 1986).

(9) *Reduced occipital squama height (lambda-inion chord) compared to nuchal length (inion-opisthion chord) (REOH)*. Ranging from the longer occipital squama height of Neanderthals, African late archaic *H. sapiens*, Skhūl-Qafzeh and modern human groups (scored 0) to the low height found in Asian *H. erectus* (scored 2). Although the condition in other hominoids would suggest that the outgroup condition should be scored 2, the predominant condition in "australopithecines" would suggest a score of 0 or 1. An outgroup score of 1 has been selected for PAUP (for discussion and data see Andrews, 1984; Hublin, 1985; Kimbel *et al.*, 1984; Rightmire, 1984; Stringer, 1984*a*; Wood, 1984).

(10) *Endocranial volume (BRN)*. Ranging from the low values of *habilis* A and B (both scored 1, although there are data suggesting that values for A were typically larger than for B) to

the high values found in the Neanderthal and Skhūl-Qafzeh groups (scored 4). There seems no doubt that the outgroup score here should be 0 (for discussion see Holloway, 1983; Stringer, 1984*b*, 1986; Trinkaus, 1983).

(11) *Dental reduction (REDN)*. This is a composite assessment of crown size reduction through the whole dentition and ranges from the large dentition of *H. habilis* A (scored 1) to the small dentition of modern humans (scored 4). There is little doubt that the outgroup score should reflect a dentition at least as large as that of *H. habilis* A, and an outgroup score of 0 has been assigned (White *et al.*, 1981; Wolpoff, 1980).

Results of PAUP analyses

Because of the relatively small size of the data set it was possible to run the powerful "branch and bound" algorithm for tree construction, and using the FARRIS option, character changes in opposite directions were equally permissible. Three equally parsimonious trees were produced with a branch length of 43 and a high consistency index of 0·744. They differed only in their resolution of the Neanderthal/African late archaic *sapiens*/Skhūl-Qafzeh and modern human relationship. While the Skhūl-Qafzeh and modern human groups were always linked as sister taxa, their sister group was Neanderthals in one solution, African late archaic *H. sapiens* in a second solution, and in a third solution there was an actual trichotomy. Representing these three solutions by an unresolved trichotomy, it was possible to produce the consensus cladogram shown in Figure 1.

As can be seen by the accompanying character assignments, the characters which were given different values at nodes resolving the NEA/AAS/S-Q/MOD relationship were postcranial robusticity (affected by whether the AAS group was assigned a postcranial robusticity of 0 or 1), midfacial projection, NLH/EKB index and brain size. In one analysis (the trichotomy), Neanderthals and African late *H. sapiens* were only linked by archaic symplesiomorphies retained from node 6 rather than by synapomorphies, while in a second analysis Neanderthals and Skhūl-Qafzeh/modern were linked by synapomorphies of increased midfacial projection (MFPJ 2, subsequently further increased to MFPJ 3 in NEA, and reversed to MFPJ 1 in S/Q) and brain size (BRN 4, subsequently reversed to BRN 3 in MOD). In a third analysis, African late archaic *H. sapiens* and Skhūl-Qafzeh/modern were linked by the synapomorphies of low postcranial robusticity (PROB 0, inferred by PAUP for AAS) and low NLH/EKB ratio (N/E 0, subsequently reversed to N/E 1 in MOD).

Three especially striking results are apparent from the consensus tree depicted in Figure 1. First, the *H. habilis* groups are plesiomorphous, while the modern *H. sapiens* group, by contrast, is highly derived. Second, the African and Asian *H. erectus* groups are linked cladistically. Third, the early archaic *H. sapiens* group is also assigned to this *H. erectus* clade. These results suggest firstly that apparent resemblances between *H. habilis* and modern *H. sapiens* are homoplasies caused by evolutionary reversals in later human evolution (see further discussion below). Secondly, that an increase in cranial flattening (CRFL) and occipital angulation (OCAN) does link early African *H. erectus* with Asian *H. erectus*. Thirdly, in the assignment of the early archaic *H sapiens* group, the synapomorphies of cranial and postcranial robusticity, supraorbital torus morphology and occipital angulation have taken precedence over the apparent *H. sapiens* synapomorphies of reduced cranial flattening, increased midfacial projection, reduced prognathism and increased brain size. PAUP has treated these supposed *H. sapiens* synapomorphies as parallelisms, with cranial flattening undergoing a reversal within the *H. erectus* clade.

The characterization of *Homo* (*versus* the outgroup character set) consists only of a reduction in facial prognathism (REPN) and dental size (REDN), and an increase in endocranial volume (BRN). Greater characterization is possible if the *H. habilis* A group is excluded, via a further reduction in prognathism and dental size, and an increase in occipital angulation (OCAN). Within *Homo*, the *H. habilis* A group is only characterized by the apparent autapomorphy of decreased midfacial projection (MFPJ 0), while *H. habilis* B shows no distinction from node 2 if assigned the same postcranial robusticity (PROB 1).

As mentioned already, the apparent similarities of the Skhūl-Qafzeh/modern groups to the *H. habilis* groups have, in most cases, been treated as reversals by PAUP. It was only in a separate series of PAUP analyses using *unordered* characters (i.e., where sequential

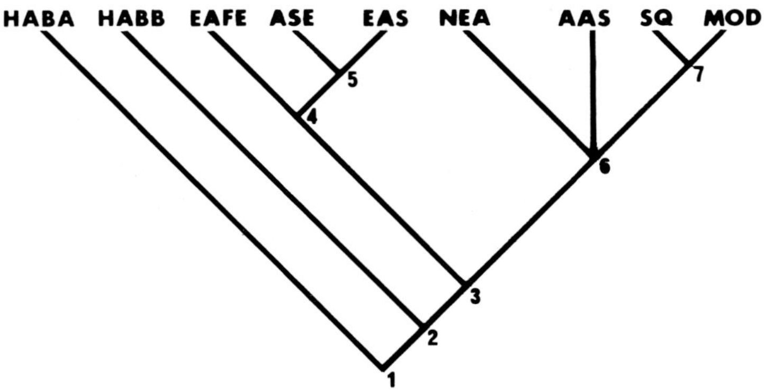

Figure 1. Consensus cladogram of three most parsimonious trees generated by PAUP (FARRIS and branch and bound options used, with characters treated as ordered). Branch length 43, consistency index 0·744.

Character assignment changes at nodes (inferred characters bracketed, outgroup score also given).

OUTGROUP. CROB 0; PROB 1; BROW 1; CRFL 1; OCAN 0; MFPJ 1; N/E 1; REPN 0; REOH 1; BRN 0; REDN 0.

1. REPN 1; BRN 1; REDN 1.
1-HABA. MFPJ O. [*H. habilis* A]
2. OCAN 1; REPN 2; REDN 2.
2-HABB. (PROB 1). [*H. habilis* B]
3. CROB 1; BROW 2; BRN 2; REDN 3.
4. CRFL 2; OCAN 2.
4-EAFE.—. [early African *H. erectus*.]
5. CROB 2; PROB 2; BROW 3; OCAN 3.
5-ASE. CROB 3; N/E 0; REOH 2. [Asian *H. erectus*.]
5 EAS. CRFL 1; MFPJ 2; REPN 3; BRN 3. [early archaic *H. sapiens*.]
6. REPN 3; REOH 0; BRN 3.
6-NEA/AAS/S-Q+MOD. Combinations of PROB 0/1; MFPJ 1/2; N/E 0/1 BRN 3/4. [Neanderthals, African late archaic *H. sapiens*, Skhūl-Qafzeh, modern humans.]
6-NEA. MFPJ 3; N/E 2. [Neanderthals.]
6-AAS.—.[African late archaic *H. sapiens*.]
7. CROB 0; PROB 0; BROW 1; CRFL 0; OCAN 0.
7-SQ.—.[Skhūl-Qafzlh.]
7-MOD. BROW 0; REPN 4; REDN 4. [Modern humans.]

changes of character state assignments were *not* preferred), where this was not the case. However, a scheme similar to that advocated by Andrews (1984) concerning the possibility of a *H. habilis*-modern *H. sapiens* sister-group relationship was still not favoured by PAUP, since all seven of the most parsimonious trees produced by PAUP using unordered characters (with lengths of 30 and consistency indices of 0·763) treated the similarities of *H. habilis* and Skhūl-Qafzeh/modern as symplesiomorphies, compared with the synapomorphies linking the other groups.

Under the user-tree option, a conventional time-based phyletic sequence of the groups was analysed by PAUP for comparison (Figure 2). This tree was certainly less parsimonious than PAUP's three best trees (branch length increased to 46, and the consistency index dropped to 0·696), but relatively this tree did not score badly. The modelling of character change approximated quite close to my own preconceived ideas, with the early African *H. erectus*, early archaic *H. sapiens* and African late archaic *H. sapiens* groups looking primitive, without autapomorphies, while the Neanderthals were characterized by autapomorphies in increased midfacial projection (MFPJ increasing from

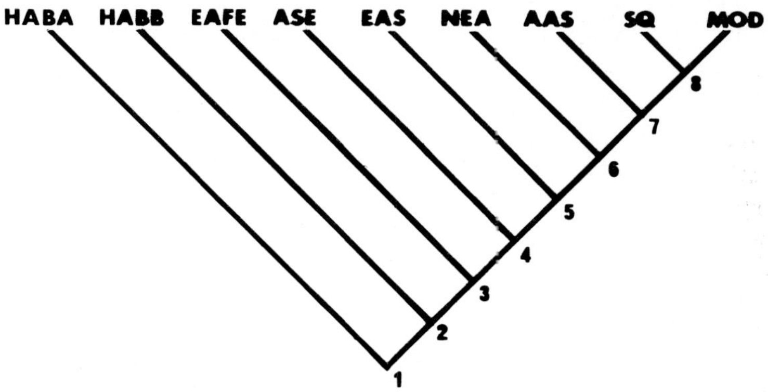

Figure 2. Cladogram entered under user-tree option for comparison with that of Figure 1. Branch length 46, consistency index 0·696.

Character assignment changes at nodes (inferred changes bracketed, outgroup score also given).

OUTGROUP. CROB 0; PROB 1; BROW 1; CRFL 1; OCAN 0; MFPJ 1; N/E 1; REPN 0; REOH 1; BRN 0; REDN 0.

1. REPN 1; BRN 1; REDN 1.
1-HABA. MFPJ 0. [*H. habilis* A.]
2. OCAN 1; REPN 2; REDN 2.
2-HABB. (PROB 1). [*H. habilis* B.]
3. CROB 1; BROW 2; CRFL 2; OCAN 2; BRN 2; REDN 3.
3-EAFE.—.[Early African *H. erectus.*]
4. CROB 2; PROB 2; BROW 3; OCAN 3.
4-ASE. CROB 3; REOH 2; N/E 0. [Asian *H. erectus.*]
5. CRFL 1; MFPJ 2; REPN 3; BRN 3.
5-EAS.—. [Early archaic *H. sapiens.*]
6. CROB 1; PROB 1; BROW 2; OCAN 1; REOH 0.
6-NEA. MFPJ 3; N/E 2; BRN 4. [Neanderthals.]
7. PROB 0; MFPJ 1; N/E 0.
7-AAS. (PROB 0). [African late archaic *H. sapiens.*]
8. CROB 0; BROW 1; CRFL 0; OCAN 0.
8-S/Q. BRN 4. [Skhūl-Qafzeh.]
8-MOD. BROW 0; MFPJ 2; N/E 1; REPN 4; REDN 4. [modern humans.]

2 to 3), NLH/EKB index (N/E from 1 to 2) and brain size (BRN from 3 to 4). Synapomorphies linking African late archaic *H. sapiens* to the Skhūl-Qafzeh/modern groups were decreased midfacial projection (MFPJ from 2 to 1), decreased NLH/EKB index (N/E from 1 to 0) and (assumed by PAUP for African late archaic *sapiens*) a decrease in postcranial robusticity (PROB from 1 to 0). The distribution of characters in this part of the tree meant that brain size increased in parallel between Neanderthals and Skhūl-Qafzeh, while midfacial projection and NLH/EKB index also increased with modern humans (a reversal of MFPJ from 1 to 2, and of NLH/EKB from 0 to 1).

Concluding remarks

Using PAUP with the small data set provided, it was possible to answer the questions posed earlier. PAUP provided three equally parsimonious best trees which differed from the less parsimonious "conventional" tree by the formation of a clade containing the early African and Asian *H. erectus* groups, together with the early archaic *sapiens* group, suggesting that the usual grouping of early and late *H. sapiens* forms is, in fact, polyphyletic (in the sense of Holmes, 1980). "Advanced" *H. sapiens*-like characters of the archaic *sapiens* group were treated as homoplastic (parallelisms) by PAUP. PAUP analyses did not resolve the Neanderthal/African late archaic *H. sapiens*/Skhūl-Qafzeh + modern *H. sapiens* relationship, so additional characters would be helpful here. Provided ordered characters were used, the *H. habilis* groups were treated as plesiomorphous by PAUP, while modern *H. sapiens* was highly derived.

This exploratory cladistic analysis using PAUP has been instructive, and it is hoped that it will be followed by a more detailed analysis using additional characters and groups. For example, the composite characters used here (such as "cranial robusticity" and "postcranial robusticity") could be separated into a number of distinct elements, and information from recently discovered partial skeletons of early African *H. erectus* (WT 15000, Kenya), "archaic *H. sapiens*" (Yinkou, China) and a Neanderthal (Kebara, Israel) should provide new data for anatomical regions such as the pelvis. Problematic specimens usually assigned to archaic *H. sapiens*, such as Steinheim and Omo-Kibish 2, would also be useful additions to the analyses, if it can be determined in which group(s) they should be placed. A larger data set should allow further testing of the phylogenetic position of the early archaic *H. sapiens* group, and a resolution of the later *H. sapiens* trichotomy. It may also be possible to provide more characterization of the *H. habilis*, early African *H. erectus* and African late archaic *H. sapiens* groups than was possible from the small data set used here.

Acknowledgements

I would like to thank Chris Humphries and Peter Forey (BM[NH]) for their valuable help in explaining the PAUP program and running the analyses, and Robert Kruszynski (BM[NH]) for preparing Figures 1 and 2. The comments of three referees were also constructive in improving this paper.

References

Andrews, P. (1984). An alternative interpretation of the characters used to define *Homo erectus*. *Cour. Forsch. Inst. Senckenberg* **69,** 167–175.

Andrews, P. & Franzen, J. L., Eds (1984). The early evolution of Man with special emphasis on southeast Asia and Africa. *Cour. Forsch. Inst. Senckenberg* **69** (whole volume).

Bonde, N. (1977). Cladistic classification as applied to vertebrates. In (M. K. Hecht, P. C. Goody & B. M. Hecht, Eds) *Major Patterns in Vertebrate Evolution*, pp. 741–804. New York: Plenum.
Bonde, N. (1981). Problems of species concepts in palaeontology. *Paper for International Symposium on Concepts and Methods in Paleontology, Barcelona, 1981*, pp. 19–34.
Bräuer, G. (1984). A craniological approach to the origin of anatomically modern *Homo sapiens* in Africa and implications for the appearance of modern Europeans. In (F. Smith & F. Spencer, Eds) *The Origins of Modern Humans*, pp. 327–410. New York: Alan Liss.
Bräuer, G. & Leakey, R. E. (1986). The ES-11693 cranium from Eliye Springs, West Turkana, Kenya. *J. hum. Evol.* **15,** 289–312.
Brown, F., Harris, J. R., Leakey, R. E. & Walker, A. (1985). Early *Homo erectus* skeleton from west Lake Turkana, Kenya. *Nature* **316,** 788–792.
Dean, M. C. (1986). *Homo* and *Paranthropus*: similarities in the cranial base and developing dentition. In (B. Wood, L. Martin & P. Andrews, Eds) *Major Topics in Primate and Human Evolution*, pp. 249–265. Cambridge: Cambridge University Press.
Delson, E. (Ed.) (1985). *Ancestors: the Hard Evidence.* New York: Alan Liss.
Delson, E., Eldredge, N. & Tattersall, I. (1977). Reconstruction of hominid phylogeny: a testable framework based on cladistic analysis. *J. hum. Evol.* **6,** 263–278.
Eldredge, N. & Tattersall, I. (1975). Evolutionary models, phylogenetic reconstruction, and another look at hominid phylogeny. *Contr. Primat.* **5,** 218–242.
Fink, W. L. (1986). Microcomputers and phylogenetic analysis. *Science* **234,** 1135–1139.
Holloway, R. L. (1983). Human brain evolution: a search for units, models and synthesis. *Can. J. Anthrop.* **3,** 215–230.
Holmes, E. B. (1980). Reconsideration of some systematic concepts and terms. *Evol. Theory* **5,** 35–87.
Howells, W. W. (1973). Cranial variation in Man. *Pap. Peabody Mus.* **67,** 1–259 (whole volume).
Howells, W. W. (1980). *Homo erectus*—who, when and where: a survey. *Yb. phys. Anthrop.* **23,** 1–23.
Hublin, J. J. (1985). Human fossils from the North African Middle Fleistocene and the origin of *Homo sapiens.* In (E. Delson, Ed.) *Ancestors: the Hard Evidence*, pp. 283–288. New York: Alan Liss.
Hublin, J. J. (1986). Some comments on the diagnostic features of *Homo erectus. Anthropos (Brno).* **23,** 175–187.
Kimbel, W. H., White, T. D. & Johanson, D. C. (1984). Cranial morphology of *Australopithecus afarensis:* a comparative study based on a composite reconstruction of the adult skull. *Am. J. phys. Anthrop.* **64,** 337–388.
Kluge, A. G. & Farris, J. S. (1969). Quantitative phyletics and the evolution of anurans. *Syst. Zool.* **18,** 1–32.
Olson, T. D. (1978). Hominid phylogenetics and the existence of *Homo* in member I of the Swartkrans Formation, South Africa. *J. hum. Evol.* **7,** 159–178.
Olson, T. D. (1985). Cranial morphology and systematics of the Hadar Formation hominids and *"Australopithecus" africanus.* In (E. Delson, Ed.) *Ancestors: the Hard Evidence*, pp. 102–119. New York: Alan Liss.
Rightmire, G. P. (1984). Comparisons of *Homo erectus* from Africa and Southeast Asia. *Cour. Forsch. Inst.Senckenberg* **69,** 83–98.
Skelton, R. R., McHenry, H. M. & Drawhorn, G. M. (1986). Phylogenetic analysis of early hominids. *Curr. Anthrop..* **27,** 21–43.
Stringer, C. B. (1983). Some further notes on the morphology and dating of the Petralona hominid. *J. hum. Evol.* **12,** 731–742.
Stringer, C. B. (1984*a*). The definition of *Homo erectus* and the existence of the species in Africa and Europe. *Cour. Forsch. Inst. Senckenberg.* **69,** 131–143.
Stringer, C. B. (1984*b*). Human evolution and biological adaptation in the Pleistocene. In (R. Foley, Ed.) *Hominid Evolution and Community Ecology*, pp. 55–83. London: Academic Press.
Stringer, C. B. (1985). Middle Pleistocene hominid variability and the origin of late Pleistocene humans. In (E. Delson, Ed.) *Ancestors: the Hard Evidence*, pp. 289–295. New York: Alan Liss.
Stringer, C. B. (1986). The credibility of *Homo habilis.* In (B. Wood, L. Martin & P. Andrews, Eds) *Major topics in Primate and Human Evolution*, pp. 266–294. Cambridge: Cambridge University Press.
Stringer, C. B. & Trinkaus, E. (1981). The Shanidar Neanderthal crania. In (C. B. Stringer, Ed.) *Aspects of Human Evolution*, pp. 129–165. London: Taylor & Francis.
Stringer, C. B., Hublin, J. J. & Vandermeersch, B. V. (1984). The origin of anatomically modern humans in western Europe. In (F. Smith & F. Spencer, Eds) *The Origins of Modern Humans*, pp. 51–135. New York: Alan Liss.
Swofford, D. L. (1985). *PAUP: version 2·4.* Champaign: Illinois Natural History Survey.
Tattersall, I. (1986). Species recognition in human paleontology. *J. hum. Evol.* **15,** 165–175.
Tattersall, I. & Eldredge, N. (1977). Fact, theory and fantasy in human paleontology. *Am. Scient.* **65,** 204–211.
Trinkaus, E. (1983). *The Shanidar Neandertals.* New York: Academic Press.
Trinkaus, E. (1984*a*). Does KNM-ER 1481A establish *Homo erectus* at 2·0 myr BP? *Am. J. phys. Anthrop.* **64,** 137–139.
Trinkaus, E. (1984*b*). Western Asia. In (F. Smith & F. Spencer, Eds) *The Origins of Modern Humans*, pp. 251–293. New York: Alan Liss.

Vandermeersch, B. V. (1981). *Les hommes fossiles de Qafzeh (Israël)*. Paris: CNRS.
White, T. D., Johanson, D. C. & Kimbel, W. H. (1981). *Australopithecus africanus:* Its Phyletic Position Reconsidered. *S. Afr. J. Sci.* **77,** 445–470.
Wiley, E. O. (1981). *Phylogenetics: the Theory and Practice of Phylogenetic Systematics*. New York: Wiley.
Wolpoff, M. H. (1980). *Paleoanthropology*. New York: Knopf.
Wood, B. (1984). The origin of *Homo erectus*. *Cour. Forsch. Inst. Senckenberg* **69,** 99–111.
Wood, B., Martin, L. & Andrews, P. (1986). *Major Topics in Primate and Human Evolution*. Cambridge: Cambridge University Press.